ON HUMAN NATURE

On Human Nature

EDWARD O. WILSON

Harvard University Press
Cambridge, Massachusetts
London, England
1978

LIBRARY OF CONGRESS CATALOGING IN PUBLICATION DATA
WILSON, EDWARD OSBORNE, 1929–
   ON HUMAN NATURE.

   BIBLIOGRAPHY: P.
   INCLUDES INDEX.
   I. SOCIOLOGY. 2. SOCIAL DARWINISM. I. TITLE.
GN365.9.W54   301.2   78–17675
ISBN 0–674–63441–1

What though these reasonings concerning human nature seem abstract and of difficult comprehension, this affords no presumption of their falsehood. On the contrary, it seems impossible that what has hitherto escaped so many wise and profound philosophers can be very obvious and easy. And whatever pains these researches may cost us, we may think ourselves sufficiently rewarded, not only in point of profit but of pleasure, if, by that means, we can make any addition to our stock of knowledge in subjects of such unspeakable importance.

Hume, *An Inquiry Concerning Human Understanding*

Contents

Preface

On Human Nature is the third book in a trilogy that unfolded without my being consciously aware of any logical sequence until it was nearly finished. The final chapter of *The Insect Societies* (1971) was entitled "The Prospect for a Unified Sociobiology." In it I suggested that the same principles of population biology and comparative zoology that have worked so well in explaining the rigid systems of the social insects could be applied point by point to vertebrate animals. In time, I said, we will account for both termite colonies and troops of rhesus monkeys by a single set of parameters and one quantitative theory. Unable to resist the rhetoric of my own challenge, I set out to learn the large and excellent literature on vertebrate social behavior and wrote *Sociobiology: The New Synthesis* (1975). In its final chapter "Man: From Sociobiology to Sociology," I argued that the biological principles which now appear to be working reasonably well for animals in general can be extended profitably to the social sciences. This suggestion created an unusual amount of interest and controversy.

The aftermath of the publication of *Sociobiology* led me to read more widely on human behavior and drew me to many seminars and written exchanges with social scientists. I became more persuaded than ever that the time has at last arrived to close the famous gap between the two cultures, and that general sociobiology, which is simply the extension of population biology and evolutionary theory to social organization, is the appropriate instrument for the effort. *On Human Nature* is an exploration of that thesis.

But this third book could not be a textbook or a conventional synthesis of the scientific literature. To address human behavior systematically is to make a potential topic of every corridor in the labyrinth of the human mind, and hence to consider not just the social sciences but also the humanities, including philosophy and the process of scientific discovery itself. Consequently, *On Human Nature* is not a work of science; it is a work about science, and about how far the natural sciences can penetrate into human behavior before they will be transformed into something new. It examines the reciprocal impact that a truly evolutionary explanation of human behavior must have on the social sciences and humanities. *On Human Nature* may be read for information about behavior and sociobiology, which I have been careful to document. But its core is a speculative essay about the profound consequences that will follow as social theory at long last meets that part of the natural sciences most relevant to it.

Opinion on the merit of these arguments will no doubt be as sharply divided as it was on the sections dealing with human behavior in *Sociobiology*. At the risk of surrendering advantage to those whose beliefs leave them no option but rejection, I wish to say the following to others who are prone to read this book uncritically as a tested product of science: I might easily be wrong—in any particular conclusion, in the grander hopes for the role of the natural sciences, and in the trust gambled on scientific materialism. This qualification does not represent false modesty but instead is an attempt to maintain strength. The uncompromising application of evolutionary theory to all aspects of human existence will come to nothing if the scientific spirit itself falters, if ideas are not constructed so as to be submitted to objective testing and hence made mortal. The social sciences are still too young and weak, and evolutionary theory itself still too imperfect, for the propositions reviewed here to be carved in stone. It is my conviction nonetheless that the existing evidence favors them and through them the broader confidence in biological inquiry that forms the main thrust of this exposition.

I have been blessed with friends and colleagues who provided enormously useful aid and advice during the preparation of the book. They of course do not agree with everything I have said, and I exonerate them all from the errors that still remain. Their names follow: Richard D. Alexander, Jerome H. Barkow, Daniel Bell, William I. Bennett, Herbert Bloch, William E. Boggs, John T. Bonner, John E. Boswell, Ralph W. Burhoe, Donald T. Campbell, Arthur Caplan, Napoleon A. Chagnon, George A. Clark, Robert K. Colwell, Bernard D. Davis, Irven DeVore, Mildred Dickeman, Robin Fox, Daniel G. Freedman, William D. Hamilton, Richard J. Herrnstein, Bert Hölldobler, Gerald Holton, Sarah Blaffer Hrdy, Harry J. Jerison, Mary-Claire King, Melvin Konner, George F. Oster, Orlando Patterson, John E. Pfeiffer, David Premack, W. V. Quine, Jon Seger, Joseph Shepher, B. F. Skinner, Frank Sulloway, Lionel Tiger, Robert L. Trivers, Pierre van den Berghe, Arthur W. Wang, James D. Weinrich, Irene K. Wilson, Richard W. Wrangham.

As she has done for my previous books, Kathleen M. Horton aided in bibliographic research and typed the successive drafts of the manuscript. Her assistance has improved the accuracy and efficiency of my work by an amount I would be afraid to try to measure.

Chapter 1 contains relatively unchanged portions of my previous articles "The Social Instinct," *Bulletin of the American Academy of Arts and Sciences*, 30: 11–24 (1976) and "Biology and the Social Sciences," *Daedalus*, 106(4): 127–140 (1977); Chapters 5 and 7 contain most of the content of "Human Decency Is Animal" (*The New York Times Magazine*, October 12, 1975); and Chapters 4 and

8 contain a few sections from Chapter 27 of *Sociobiology*. The permission of the publishers to reproduce this material is appreciated. Permission for the quotation of work by other authors has been obtained variously from the University of California Press, the University of Chicago Press, and Macmillan Company; the specific citations are given in the bibliographic notes.

Chapter 1. Dilemma

These are the central questions that the great philosopher David Hume said are of unspeakable importance: How does the mind work, and beyond that why does it work in such a way and not another, and from these two considerations together, what is man's ultimate nature?

We keep returning to the subject with a sense of hesitancy and even dread. For if the brain is a machine of ten billion nerve cells and the mind can somehow be explained as the summed activity of a finite number of chemical and electrical reactions, boundaries limit the human prospect — we are biological and our souls cannot fly free. If humankind evolved by Darwinian natural selection, genetic chance and environmental necessity, not God, made the species. Deity can still be sought in the origin of the ultimate units of matter, in quarks and electron shells (Hans Küng was right to ask atheists why there is something instead of nothing) but not in the origin of species. However much we embellish that stark conclusion with metaphor and imagery, it remains the philosophical legacy of the last century of scientific research.

No way appears around this admittedly unappealing proposition. It is the essential first hypothesis for any serious consideration of the human condition. Without it the humanities and social sciences are the limited descriptors of surface phenomena, like astronomy without physics, biology without chemistry, and mathematics without algebra. With it, human nature can be laid open as an object of fully empirical research, biology can be put to the service of liberal education, and our self-conception can be enormously and truthfully enriched.

But to the extent that the new naturalism is true, its pursuit seems certain to generate two great spiritual dilemmas. The first is that no species, ours included, possesses a purpose beyond the imperatives created by its genetic history. Species may have vast potential for material and mental progress but they lack any immanent purpose or guidance from agents beyond their immediate environment or even an evolutionary goal toward which their molecular architecture automatically steers them. I believe that the human mind is constructed in a way that locks it inside this fundamental constraint and forces it to make choices with a purely biological instrument. If the brain evolved by natural selection, even the capacities to select particular esthetic judgments and religious beliefs must have arisen by the same mechanistic process. They are either direct adaptations to past environments in which the ancestral human populations evolved or at most constructions thrown up secondarily by deeper, less visible activities that were once adaptive in this stricter, biological sense.

The essence of the argument, then, is that the brain exists because it promotes the survival and multiplication of the genes that direct its assembly. The human mind is a device for survival and reproduction, and reason is just one of its various techniques. Steven Weinberg has pointed out that physical reality remains so mysterious even to physicists because of the extreme improbability that it was constructed to be understood by the human mind. We can reverse that insight to note with still greater force that the intellect was not constructed to understand atoms or even to understand itself but to promote the survival of human genes. The reflective person knows that his life is in some incomprehensible manner guided through a biological ontogeny, a more or less fixed order of life stages. He senses that with all the drive, wit, love, pride, anger, hope, and anxiety that characterize the species he will in the end be sure only of helping to perpetuate the same cycle. Poets have defined this truth as tragedy. Yeats called it the coming of wisdom:

> Though leaves are many, the root is one;
> Through all the lying days of my youth
> I swayed my leaves and flowers in the sun;
> Now I may wither into the truth.

The first dilemma, in a word, is that we have no particular place to go. The species lacks any goal external to its own biological nature. It could be that in the next hundred years humankind will thread the needles of technology and politics, solve the energy and materials crises, avert nuclear war, and control reproduction. The world can at least hope for a stable ecosystem and a well-nourished population. But what then? Educated people everywhere like to believe that beyond material needs lie fulfillment and the realization of individual potential. But what is fulfillment, and to what ends may potential be realized? Traditional religious beliefs have been eroded, not so much by humiliating disproofs of their mythologies as by the growing awareness that beliefs are really enabling mechanisms for survival. Religions, like other human institutions, evolve so as to enhance the persistence and influence of their practitioners. Marxism and other secular religions offer little more than promises of material welfare and a legislated escape from the consequences of human nature. They, too, are energized by the goal of collective self-aggrandizement. The French political observer Alain Peyrefitte once said admiringly of Mao Tse-tung that "the Chinese knew the narcissistic joy of loving themselves in him. It is only natural that he should have loved himself through them." Thus does ideology bow to its hidden masters the genes, and the highest impulses seem upon closer examination to be metamorphosed into biological activity.

The more somber social interpreters of our time, such as Robert Heilbroner, Robert Nisbet, and L. S. Stavrianos, perceive Western civilization and ultimately mankind as a whole to be in immediate danger of decline. Their reasoning leads easily to a vision of post-ideological societies whose members will regress steadily toward self-indulgence. "The will to power will not have vanished entirely," Gunther Stent writes in *The Coming of the Golden Age*,

> but the distribution of its intensity will have been drastically altered. At one end of this distribution will be the minority of the people whose work will keep intact the technology that sustains the multitude at a high standard of living. In the middle of the distribution will be found a type, largely unemployed, for whom the distinction between the real and the illusory will still be meaningful . . . He will retain interest in the world and seek satisfaction from sensual pleasures. At the other end of the spectrum will be a type largely unemployable, for whom the boundary of the real and the imagined will have been largely dissolved, at least to the extent compatible with his physical survival.

Thus the danger implicit in the first dilemma is the rapid dissolution of transcendental goals toward which societies can organize their energies. Those goals, the true moral equivalents of war, have faded; they went one by one, like mirages, as we drew closer. In order to search for a new morality based upon a more truthful definition of man, it is necessary to look inward, to dissect the machinery of the mind and to retrace its evolutionary history. But that effort, I predict, will uncover the second dilemma, which is the choice that must be made among the ethical premises inherent in man's biological nature.

At this point let me state in briefest terms the basis of the second dilemma, while I defer its supporting argument to the next chapter: innate censors and motivators exist in the brain that deeply and unconsciously affect our ethical premises; from these roots, morality evolved as instinct. If that perception is correct, science may soon be in a position to investigate the very origin and meaning of human values, from which all ethical pronouncements and much of political practice flow.

Philosophers themselves, most of whom lack an evolutionary perspective, have not devoted much time to the problem. They examine the precepts of ethical systems with reference to their consequences and not their origins. Thus John Rawls opens his influential *A Theory of Justice* (1971) with a proposition he regards as beyond dispute: "In a just society the liberties of equal citizenship are taken as settled; the rights secured by justice are not subject to political bargaining or to the calculus of social interests." Robert Nozick begins *Anarchy, State, and Utopia* (1974) with an equally firm proposition: "Individuals have rights, and there are things no person or group may do to them (without violating their rights). So strong and far-reaching are these rights they raise the question of what, if anything, the state and its officials.may do." These two premises are somewhat different in content, and they lead to radically different prescriptions. Rawls would allow rigid social control to secure as close an approach as possible to the equal distribution of society's rewards. Nozick sees the ideal society as one governed by a minimal state, empowered only to protect its citizens from force and fraud, and with unequal distribution of rewards wholly permissible. Rawls rejects the meritocracy; Nozick accepts it as desirable except in those cases where local communities voluntarily decide to experiment with egalitarianism. Like everyone else, philosophers measure their personal emotional responses to various alternatives as though consulting a hidden oracle.

That oracle resides in the deep emotional centers of the brain, most probably within the limbic system, a complex array of neurons and hormone-secreting cells located just beneath the "thinking" portion of the cerebral cortex. Human emotional responses and the more general ethical practices based on them have been programmed to a substantial degree by natural selection over thousands of generations. The challenge to science is to measure the tightness of the constraints caused by the programming, to find their source in the brain, and to decode their significance through the reconstruction of the evolutionary history of the mind. This enterprise will be the logical complement of the continued study of cultural evolution.

Success will generate the second dilemma, which can be stated as follows: Which of the censors and motivators should be obeyed and which ones might better be curtailed or sublimated? These guides are the very core of our humanity. They and not the belief in spiritual apartness distinguish us from electronic computers. At some time in the future we will have to decide how human we wish to remain—in this ultimate, biological sense—because we must consciously choose among the alternative emotional guides we have inherited. To chart our destiny means that we must shift from automatic control based on our biological properties to precise steering based on biological knowledge.

Because the guides of human nature must be examined with a complicated arrangement of mirrors, they are a deceptive subject, always the philosopher's deadfall. The only way forward is to study human nature as part of the natural sciences, in an attempt to integrate the natural sciences with the social sciences and humanities. I can conceive of no ideological or formalistic shortcut. Neurobiology cannot be learned at the feet of a guru. The consequences of genetic history cannot be chosen by legislatures. Above all, for our own physical well-being if nothing else, ethical philosophy must not be left in the hands of the merely wise. Although human progress can be achieved by intuition and force of will, only hard-won empirical knowledge of our biological nature will allow us to make optimum choices among the competing criteria of progress.

The important initial development in this analysis will be the conjunction of biology and the various social sciences—psychology, anthropology, sociology, and economics. The two cultures have only recently come into full sight of one another. The result has been a predictable mixture of aversions, misunderstandings, overenthusiasm, local conflicts, and treaties. The situation can be summarized by saying that biology stands today as the antidiscipline of the social sciences. By the word "antidiscipline" I wish to emphasize the special adversary relation that often exists when fields of study at adjacent levels of organization first begin to interact. For chemistry there is the antidiscipline of many-body physics; for molecular biology, chemistry; for physiology, molecular biology; and so on upward through the paired levels of increasing specification and complexity.

In the typical early history of a discipline, its practitioners believe in the novelty and uniqueness of their subject. They devote lifetimes to special entities and patterns and during the early period of exploration they doubt that these phenomena can be reduced to simple laws. Members of the antidiscipline have a different attitude. Having chosen as their primary subject the units of the lower level of organization, say atoms as opposed to molecules, they believe that the next discipline above can and must be reformulated by their own laws: chemistry by the laws of physics, biology by the laws of chemistry, and so on downward. Their interest is relatively narrow, abstract, and exploitative. P.A.M. Dirac, speaking of the theory of the hydrogen atom, could say that its consequences would unfold as mere chemistry. A few biochemists are still content in the belief that life is "no more" than the actions of atoms and molecules.

It it easy to see why each scientific discipline is also an antidiscipline. An adversary relationship is probable because the devotees of the two adjacent organizational levels—such as atoms versus molecules—are initially committed to their own methods and ideas when they focus on the upper level (in this case, molecules). By today's standards a broad scientist can be defined as one who is a student of three subjects: his discipline (chemistry in the example cited), the lower antidiscipline (physics), and the subject to which his specialty stands as antidiscipline (the chemical aspects of biology). A well-rounded expert on the nervous system, to take a second, more finely graded example, is deeply versed in the structure of single nerve cells, but he also understands the chemical basis of the impulses that pass through and between these cells, and he hopes to explain how nerve cells work together to produce elementary patterns of behavior. Every successful scientist treats differently each of the three levels of phenomena surrounding his specialty.

The interplay between adjacent fields is tense and creative at the beginning, but with the passage of time it becomes fully complementary. Consider the origins of molecular biology. In the late 1800s the microscopic study of cells (cytology) and the study of chemical processes within and around the cells (biochemistry) grew at an accelerating pace. Their relationship during this period was complicated, but it broadly fits the historical schema I have described. The cytologists were excited by the mounting evidence of an intricate cell architecture. They had interpreted the mysterious choreography of the chromosomes during cell division and thus set the stage for the emergence of modern genetics and experimental developmental biology. Many biochemists, on the other hand, remained skeptical of the idea that so much structure exists at the microscopic level. They thought that the cytologists were describing artifacts created by laboratory methods of fixing and staining cells for microscopic examination. Their interest lay in the more "fundamental" issues of the chemical nature of protoplasm, especially the newly formulated theory that life is based on enzymes. The cytologists responded with scorn to any notion that the cell is a "bag of enzymes."

In general, biochemists judged the cytologists to be too ignorant of chemistry to grasp the fundamental processes, while the cytologists considered the methods of the chemists inappropriate for the idiosyncratic structures of the living cell. The revival of Mendelian genetics in 1900 and the subsequent illumination of the roles of the chromosomes and genes did little at first to force a synthesis. Biochemists, seeing no immediate way to explain classical genetics, by and large ignored it.

Both sides were essentially correct. Biochemistry has now explained so much of the cellular machinery on its own terms as to justify its most extravagant early claims. But in achieving this feat, mostly since 1950, it was partially transformed into the new discipline of molecular biology, which can be defined as biochemistry that also accounts for the particular spatial arrangements of such molecules as the DNA helix and enzyme proteins. Cytology forced the development of a special kind of chemistry and the use of a battery of powerful new techniques, including electrophoresis, chromatography, density-gradient centrifugation, and x-ray crystallography. At the same time cytology metamorphosed into modern cell biology. Aided by the electron microscope, which magnifies objects by hundreds of thousands of times, it has converged in perspective and language toward molecular biology. Finally, classical genetics, by switching from fruit flies and mice to bacteria and viruses, has incorporated biochemistry to become molecular genetics.

Progress over a large part of biology has been fueled by competition among the various perspectives and techniques derived from cell biology and biochemistry, the discipline and its antidiscipline. The interplay has been a triumph for scientific materialism. It has vastly enriched our understanding of the nature of life and created materials for literature more powerful than any imagery of prescientific culture.

I suggest that we are about to repeat this cycle in the blending of biology and the social sciences and that as a consequence the two cultures of Western intellectual life will be joined at last. Biology has traditionally affected the social sciences only indirectly through technological manifestations, such as the benefits of medicine, the mixed blessings of gene splicing and other techniques of genetics, and the specter of population growth. Although of great practical importance, these matters are trivial with reference to the conceptual foundation of the social sciences. The conventional treatments of "social biology" and "social issues of biology" in our colleges and universities present some formidable intellectual challenges, but they are not addressed to the core of social theory. This core is the deep structure of human nature, an essentially biological phenomenon that is also the primary focus of the humanities.

It is all too easy to be seduced by the opposing view: that science is competent to generate only a few classes of information, that its cold, clear Apollonian method will never be relevant to the full Dionysian life of the mind, that single-minded devotion to science is dehumanizing. Expressing the mood of the counterculture, Theodore Roszak suggested a map of the mind "as a spectrum of possibilities, all of which properly blend into one another . . . At one end, we have the hard, bright lights of science; here we find information. In the center we have the sensuous hues of art; here we find the aesthetic shape of the world. At the far end, we have the dark, shadowy tones of religious experience, shading off into wave lengths beyond all perception; here we find meaning."

No, here we find obscurantism! And a curious underestimate of what the mind can accomplish. The sensuous hues and dark tones have been produced by the genetic evolution of our nervous and sensory tissues; to treat them as other than objects of biological inquiry is simply to aim too low.

The heart of the scientific method is the reduction of perceived phenomena to fundamental, testable principles. The elegance, we can fairly say the beauty, of any particular scientific generalization is measured by its simplicity relative to the number of phenomena it can explain. Ernst Mach, a physicist and forerunner of the logical positivists, captured the idea with a definition: "Science may be regarded as a minimal problem consisting of the completest presentation of facts with the least possible expenditure of thought."

Although Mach's perception has an undeniable charm, raw reduction is only half of the scientific process. The remainder consists of the reconstruction of complexity by an expanding synthesis under the control of laws newly demonstrated by analysis. This reconstitution reveals the existence of novel, emergent phenomena. When the observer shifts his attention from one level of organization to the next, as from physics to chemistry or from chemistry to biology, he expects to find obedience to all the laws of the levels below. But to reconstitute the upper levels of organization requires specifying the arrangement of the lower units and this in turn generates richness and the basis of new and unexpected principles. The specification consists of particular combinations of units, as well as particular spatial arrangements and histories of the ensembles of these elements. Consider the following simple example from chemistry. The ammonia molecule consists of a negatively charged nitrogen atom bonded to a triangle of three positively charged hydrogen atoms. If the atoms were locked in one position the ammonia molecule would have an opposite charge at each end (a dipole moment) in apparent contradiction to the symmetry laws of nuclear physics. Yet the molecule manages to behave properly: it neutralizes its dipole moment by passing the nitrogen atom back and forth through the triangle of hydrogen atoms at a frequency of thirty billion times per second. However, such symmetry is absent in the case of sugar and other large organic molecules, which are too large and complex in structure to invert themselves. They break but do not repeal the laws of physics. This specification may not be greatly interesting to nuclear physicists, but its consequences redound throughout organic chemistry and biology.

Consider a second example, closer to our subject, from the evolution of social life in the insects. In the Mesozoic Era, about 150 million years ago, primitive wasps evolved the sex-determining trait of haplodiploidy, in which fertilized eggs produced females and those left unfertilized produced males. This simple method of control may have been a specific adaptation that permitted females to choose the sex of their offspring according to the nature of the prey insects they were able to subdue. In particular, smaller prey might have been assigned to the male offspring, which require less protein in their development. But whatever its initial cause, haplodiploidy represented an evolutionary event that quite accidentally predisposed these insects to develop advanced forms of social life. The reason is that haplodiploidy causes sisters to be more closely related to each other than mothers are to daughters, and so females may derive genetic profit from becoming a sterile caste specialized for the rearing of sisters. Sterile castes engaged in rearing siblings are the essential feature of social organization in the insects. Because of its link to haplodiploidy, insect social life is almost limited to the wasps and their close relatives among the bees and ants. Furthermore, most cases can be classified either as matriarchies, in which queens control colonies of daughters, or as sisterhoods, in which sterile daughters control the egg-laying mothers. The societies of wasps, bees, and ants have proved so successful that they dominate and alter most of the land habitats of the Earth. In the forests of Brazil, their assembled forces constitute more than 20 percent of the weight of all land animals, including nematode worms, toucans, and jaguars. Who could have guessed all this from a knowledge of haplodiploidy?

Reduction is the traditional instrument of scientific analysis, but it is feared and resented. If human behavior can be reduced and determined to any considerable degree by the laws of biology, then mankind might appear to be less than unique and to that extent dehumanized. Few social scientists and scholars in the humanities are prepared to enter such a conspiracy, let alone surrender any of their territory. But this perception, which equates the method of reduction with the philosophy of diminution, is entirely in error. The laws of a subject are necessary to the discipline above it, they challenge and force a mentally more efficient restructuring, but they are not sufficient for the purposes of the discipline. Biology is the key to human nature, and social scientists cannot afford to ignore its rapidly tightening principles. But the social sciences are potentially far richer in content. Eventually they will absorb the relevant ideas of biology and go on to beggar them. The proper study of man is, for reasons that now transcend anthropocentrism, man.

Chapter 2. Heredity

We live on a planet of staggering organic diversity. Since Carolus Linnaeus began the process of formal classification in 1758, zoologists have catalogued about one million species of animals and given each a scientific name, a few paragraphs in a technical journal, and a small space on the shelves of one museum or another around the world. Yet despite this prodigious effort, the process of discovery has hardly begun. In 1976 a specimen of an unknown form of giant shark, fourteen feet long and weighing sixteen hundred pounds, was captured when it tried to swallow the stabilizing anchor of a United States Naval vessel near Hawaii. About the same time entomologists found an entirely new category of parasitic flies that resemble large reddish spiders and live exclusively in the nests of the native bats of New Zealand. Each year museum curators sort out thousands of new kinds of insects, copepods, wireworms, echinoderms, priapulids, pauropods, hypermastigotes, and other creatures collected on expeditions around the world. Projections based on intensive surveys of selected habitats indicate that the total number of animal species is between three and ten million. Biology, as the naturalist Howard Evans expressed it in the title of a recent book, is the study of life "on a little known planet."

Thousands of these species are highly social. The most advanced among them constitute what I have called the three pinnacles of social evolution in animals: the corals, bryozoans, and other colony-forming invertebrates; the social insects, including ants, wasps, bees, and termites; and the social fish, birds, and mammals. The communal beings of the three pinnacles are among the principal objects of the new discipline of sociobiology, defined as the systematic study of the biological basis of all forms of social behavior, in all kinds of organisms, including man. The enterprise has old roots. Much of its basic information and some of its most vital ideas have come from ethology, the study of whole patterns of behavior of organisms under natural conditions. Ethology was pioneered by Julian Huxley, Karl von Frisch, Konrad Lorenz, Nikolaas Tinbergen, and a few others and is now being pursued by a large new generation of innovative and productive investigators. It has remained most concerned with the particularity of the behavior patterns shown by each species, the ways these patterns adapt animals to the special challenges of their environments, and the steps by which one pattern gives rise to another as the species themselves undergo genetic evolution. Increasingly, modern ethology is being linked to studies of the nervous system and the effects of hormones on behavior. Its investigators have become deeply involved with developmental processes and even learning, formerly the nearly exclusive domain of psychology, and they have begun to include man among the species most closely scrutinized. The emphasis of ethology remains on the individual organism and the physiology of organisms.

Sociobiology, in contrast, is a more explicitly hybrid discipline that incorporates knowledge from ethology (the naturalistic study of whole patterns of behavior), ecology (the study of the relationships of organisms to their environment), and genetics in order to derive general principles concerning the biological properties of entire societies. What is truly new about sociobiology is the way it has extracted the most important facts about social organization from their traditional matrix of ethology and psychology and reassembled them on a foundation of ecology and genetics studied at the population level in order to show how social groups adapt to the environment by evolution. Only within the past few years have ecology and genetics themselves become sophisticated and strong enough to provide such a foundation.

Sociobiology is a subject based largely on comparisons of social species. Each living form can be viewed as an evolutionary experiment, a product of millions of years of interaction between genes and environment. By examining many such experiments closely, we have begun to construct and test the first general principles of genetic social evolution. It is now within our reach to apply this broad knowledge to the study of human beings.

Sociobiologists consider man as though seen through the front end of a telescope, at a greater than usual distance and temporarily diminished in size, in order to view him simultaneously with an array of other social experiments. They attempt to place humankind in its proper place in a catalog of the social species on Earth. They agree with Rousseau that "One needs to look near at hand in order to study men, but to study man one must look from afar."

This macroscopic view has certain advantages over the traditional anthropocentrism of the social sciences. In fact, no intellectual vice is more crippling than defiantly self-indulgent anthropocentrism. I am reminded of the clever way Robert Nozick makes this point when he constructs an argument in favor of vegetarianism. Human beings, he notes, justify the eating of meat on the grounds that the animals we kill are too far below us in sensitivity and intelligence to bear comparison. It follows that if representatives of a truly superior extraterrestrial species were to visit Earth and apply the same criterion, they could proceed to eat us in good conscience. By the same token, scientists among these aliens might find human beings uninteresting, our intelligence weak, our passions unsurprising, our social organization of a kind already frequently encountered on other planets. To our chagrin they might then focus on the ants, because these little creatures, with their haplodiploid form of sex determination and bizarre female caste systems, are the truly novel productions of the Earth with reference to the Galaxy. We can imagine the log declaring, "A scientific breakthrough has occurred; we have finally discovered haplodiploid social organisms in the one- to ten-millimeter range." Then the visitors might inflict the ultimate indignity: in order to be sure they had not underestimated us, they would simulate human beings in the laboratory. Like chemists testing the structural characterization of a problematic organic compound by assembling it from simpler components, the alien biologists would need to synthesize a hominoid or two.

This scenario from science fiction has implications for the definition of man. The impressive recent advances by computer scientists in the design of artificial intelligence suggests the following test of humanity: that which behaves like man *is* man. Human behavior is something that can be defined with fair precision, because the evolutionary pathways open to it have not all been equally negotiable. Evolution has not made culture all-powerful. It is a misconception among many of the more traditional Marxists, some learning theorists, and a still surprising proportion of anthropologists and sociologists that social behavior can be shaped into virtually any form. Ultra-environmentalists start with the premise that man is the creation of his own culture: "culture makes man," the formula might go, "makes culture makes man." Theirs is only a half truth. Each person is molded by an interaction of his environment, especially his cultural environment, with the genes that affect social behavior. Although the hundreds of the world's cultures seem enormously variable to those of us who stand in their midst, all versions of human social behavior together form only a tiny fraction of the realized organizations of social species on this planet and a still smaller fraction of those that can be readily imagined with the aid of sociobiological theory.

The question of interest is no longer whether human social behavior is genetically determined; it is to what extent. The accumulated evidence for a large hereditary component is more detailed and compelling than most persons, including even geneticists, realize. I will go further: it already is decisive.

That being said, let me provide an exact definition of a genetically determined trait. It is a trait that differs from other traits at least in part as a result of the presence of one or more distinctive genes. The important point is that the objective estimate of genetic influence requires comparison of two or more states of the same feature. To say that blue eyes are inherited is not meaningful without further qualification, because blue eyes are the product of an interaction between genes and the largely physiological environment that brought final coloration to the irises. But to say that the *difference* between blue and brown eyes is based wholly or partly on differences in genes is a meaningful statement because it can be tested and translated into the laws of genetics. Additional information is then sought: What are the eye colors of the parents, siblings, children, and more distant relatives? These data are compared to the very simplest model of Mendelian heredity, which, based on our understanding of cell multiplication and sexual reproduction, entails the action of only two genes. If the data fit, the differences are interpreted as being based on two genes. If not, increasingly complicated schemes are applied. Progressively larger numbers of genes and more complicated modes of interaction are assumed until a reasonably close fit can be made. In the example just cited, the main differences between blue and brown eyes are in fact based on two genes, although complicated modifications exist that make them less than an ideal textbook example. In the case of the most complex traits, hundreds of genes are sometimes involved, and their degree of influence can ordinarily be measured only crudely and with the aid of sophisticated mathematical techniques. Nevertheless, when the analysis is properly performed it leaves little doubt as to the presence and approximate magnitude of the genetic influence.

Human social behavior can be evaluated in essentially the same way, first by comparison with the behavior of other species and then, with far greater difficulty and ambiguity, by studies of variation among and within human populations. The picture of genetic determinism emerges most sharply when we compare selected major categories of animals with the human species. Certain general human traits are shared with a majority of the great apes and monkeys of Africa and Asia, which on grounds of anatomy and biochemistry are our closest living evolutionary relatives:

● Our intimate social groupings contain on the order of ten to one hundred adults, never just two, as in most birds and marmosets, or up to thousands, as in many kinds of fishes and insects.

● Males are larger than females. This is a characteristic of considerable significance within the Old World monkeys and apes and many other kinds of mammals. The average number of females consorting with successful males closely corresponds to the size gap between males and females when many species are considered together. The rule makes sense: the greater the competition among males for females, the greater the advantage of large size and the less influential are any disadvantages accruing to bigness. Men are not very much larger than women; we are similar to chimpanzees in this regard. When the sexual size difference in human beings is plotted on the curve based on other kinds of mammals, the predicted average number of females per successful male turns out to be greater than one but less than three. The prediction is close to reality; we know we are a mildly polygynous species.

● The young are molded by a long period of social training, first by closest associations with the mother, then to an increasing degree with other children of the same age and sex.

● Social play is a strongly developed activity featuring role practice, mock aggression, sex practice, and exploration.

These and other properties together identify the taxonomic group consisting of Old World monkeys, the great apes, and human beings. It is inconceivable that human beings could be socialized into the radically different repertories of other groups such as fishes, birds, antelopes, or rodents. Human beings might self-consciously *imitate* such arrangements, but it would be a fiction played out on a stage, would run counter to deep emotional responses and have no chance of persisting through as much as a single generation. To adopt with serious intent, even in broad outline, the social system of a nonprimate species would be insanity in the literal sense. Personalities would quickly dissolve, relationships disintegrate, and reproduction cease.

At the next, finer level of classification, our species is distinct from the Old World monkeys and apes in ways that can be explained only as a result of a unique set of human genes. Of course, that is a point quickly conceded by even the most ardent environmentalists. They are willing to agree with the great geneticist Theodosius Dobzhansky that "in a sense, human genes have surrendered their primacy in human evolution to an entirely new, nonbiological or superorganic agent, culture. However, it should not be forgotten that this agent is entirely dependent on the human genotype." But the matter is much deeper and more interesting than that. There are social traits occurring through all cultures which upon close examination are as diagnostic of mankind as are distinguishing characteristics of other animal species—as true to the human type, say, as wing tessellation is to a fritillary butterfly or a complicated spring melody to a wood thrush. In 1945 the American anthropologist George P. Murdock listed the following characteristics that have been recorded in every culture known to history and ethnography:

Age-grading, athletic sports, bodily adornment, calendar, cleanliness training, community organization, cooking, cooperative
labor, cosmology, courtship, dancing, decorative art, divination,
division of labor, dream interpretation, education, eschatology,
ethics, ethnobotany, etiquette, faith healing, family feasting,
fire making, folklore, food taboos, funeral rites, games, gestures,
gift giving, government, greetings, hair styles, hospitality, housing, hygiene, incest taboos, inheritance rules, joking, kin groups,
kinship nomenclature, language, law, luck superstitions, magic,
marriage, mealtimes, medicine, obstetrics, penal sanctions, personal names, population policy, postnatal care, pregnancy
usages, property rights, propitiation of supernatural beings, pu-
berty customs, religious ritual, residence rules, sexual restric-
tions, soul concepts, status differentiation, surgery, tool making,
trade, visiting, weaving, and weather control.

Few of these unifying properties can be interpreted as the inevitable outcome of either advanced social life or high intelligence. It is
easy to imagine nonhuman societies whose members are even more
intelligent and complexly organized than ourselves, yet lack a majority of the qualities just listed. Consider the possibilities inherent
in the insect societies. The sterile workers are already more cooperative and altruistic than people and they have a more pronounced
tendency toward caste systems and division of labor. If ants were
to be endowed in addition with rationalizing brains equal to our
own, they could be our peers. Their societies would display the following peculiarities:

Age-grading, antennal rites, body licking, calendar, cannibalism, caste determination, caste laws, colony-foundation rules,
colony organization, cleanliness training, communal nurseries,
cooperative labor, cosmology, courtship, division of labor, drone
control, education, eschatology, ethics, etiquette, euthanasia,

fire making, food taboos, gift giving, government, greetings, grooming rituals, hospitality, housing, hygiene, incest taboos, language, larval care, law, medicine, metamorphosis rites, mutual regurgitation, nursing castes, nuptial flights, nutrient eggs, population policy, queen obeisance, residence rules, sex determination, soldier castes, sisterhoods, status differentiation, sterile workers, surgery, symbiont care, tool making, trade, visiting, weather control,

and still other activities so alien as to make mere description by our language difficult. If in addition they were programmed to eliminate strife between colonies and to conserve the natural environment they would have greater staying power than people, and in a broad sense theirs would be the higher morality.

Civilization is not intrinsically limited to hominoids. Only by accident was it linked to the anatomy of bare-skinned, bipedal mammals and the peculiar qualities of human nature.

Freud said that God has been guilty of a shoddy and uneven piece of work. That is true to a degree greater than he intended: human nature is just one hodgepodge out of many conceivable. Yet if even a small fraction of the diagnostic human traits were stripped away, the result would probably be a disabling chaos. Human beings could not bear to simulate the behavior of even our closest relatives among the Old World primates. If by perverse mutual agreement a human group attempted to imitate in detail the distinctive social arrangements of chimpanzees or gorillas, their effort would soon collapse and they would revert to fully human behavior.

It is also interesting to speculate that if people were somehow raised from birth in an environment devoid of most cultural influence, they would construct basic elements of human social life *ab initio*. In short time new elements of language would be invented and their culture enriched. Robin Fox, an anthropologist and pioneer in human sociobiology, has expressed this hypothesis in its strongest possible terms. Suppose, he conjectured, that we performed the cruel experiment linked in legend to the Pharaoh Psammetichus and King James IV of Scotland, who were said to have reared children by remote control, in total social isolation from their elders. Would the children learn to speak to one another?

I do not doubt that they *could* speak and that, theoretically, given time, they or their offspring would invent and develop a language despite their never having been taught one. Furthermore, this language, although totally different from any known to us, would be analyzable to linguists on the same basis as other languages and translatable into all known languages. But I would push this further. If our new Adam and Eve could survive and breed — still in total isolation from any cultural influences — then eventually they would produce a society which would have laws about property, rules about incest and marriage, customs of taboo and avoidance, methods of settling disputes with a minimum of bloodshed, beliefs about the supernatural and practices relating to it, a system of social status and methods of indicating it, initiation ceremonies for young men, courtship practices including the adornment of females, systems of symbolic body adornment generally, certain activities and associations set aside for men from which women were excluded, gambling of some kind, a tool- and weapon-making industry, myths and legends, dancing, adultery, and various doses of homicide, suicide, homosexuality, schizophrenia, psychosis and neuroses, and various practitioners to take advantage of or cure these, depending on how they are viewed.

Not only are the basic features of human social behavior stubbornly idiosyncratic, but to the limited extent that they can be compared with those of animals they resemble most of all the repertories of other mammals and especially other primates. A few of the signals used to organize the behavior can be logically derived from the ancestral modes still shown by the Old World monkeys and great apes. The grimace of fear, the smile, and even laughter have parallels in the facial expressions of chimpanzees. This broad similarity is precisely the pattern to be expected if the human species descended from Old World primate ancestors, a demonstrable fact, and if the development of human social behavior retains even a small degree of genetic constraint, the broader hypothesis now under consideration.

The status of the chimpanzee deserves especially close attention. Our growing knowledge of these most intelligent apes has come to erode to a large extent the venerable dogma of the uniqueness of man. Chimpanzees are first of all remarkably similar to human beings in anatomical and physiological details. It also turns out that they are very close at the molecular level. The biochemists Mary-Claire King and Allan C. Wilson have compared the proteins encoded by genes at forty-four loci. They found the summed differences between the two species to be equivalent to the genetic distance separating nearly indistinguishable species of fruit flies, and only twenty-five to sixty times greater than that between Caucasian, Black African, and Japanese populations. The chimpanzee and human lines might have split as recently as twenty million years ago, a relatively short span in evolutionary time.

By strictly human criteria chimpanzees are mentally retarded to an intermediate degree. Their brains are only one-third as large as our own, and their larynx is constructed in the primitive ape form that prevents them from articulating human speech. Yet individuals can be taught to communicate with their human helpers by means of American sign language or the fastening of plastic symbols in sequences on display boards. The brightest among them can learn vocabularies of two-hundred English words and elementary rules of syntax, allowing them to invent such sentences as "Mary gives me apple" and "Lucy tickle Roger." Lana, a female trained by Beatrice and Robert Gardner at the University of Nevada, ordered her trainer from the room in a fit of pique by signalling, "You green shit." Sarah, a female trained by David Premack, memorized twenty-five hundred sentences and used many of them. Such well educated chimps understand instructions as complicated as "If red on green (and not vice versa) then you take red (and not green)" and "You insert banana in pail, apple in dish." They have invented new expressions such as "water bird" for duck and "drink fruit" for watermelon, essentially the same as those hit upon by the inventors of the English language.

Chimpanzees do not remotely approach the human child in the inventiveness and drive of their language. Evidence of true linguistic novelty is, moreover, lacking: no chimp genius has accomplished the equivalent of joining the sentences "Mary gives me apple" and "I like Mary" into the more complex proposition "Mary's giving me apple is why I like her." The human intellect is vastly more powerful than that of the chimpanzee. But the capacity to communicate by symbols and syntax does lie within the ape's grasp. Many zoologists now doubt the existence of an unbridgeable linguistic chasm between animals and man. It is no longer possible to say, as the leading anthropologist Leslie White did in 1949, that human behavior is symbolic behavior and symbolic behavior is human behavior.

Another chasm newly bridged is self-awareness. When Gordon G. Gallup, a psychologist, allowed chimps to peer into mirrors for two or three days, they changed from treating their reflection as a stranger to recognizing it as themselves. At this point they began to use the mirrors to explore previously inaccessible parts of their own bodies. They made faces, picked bits of food from their teeth, and blew bubbles through their pursed lips. No such behavior has ever been elicited from monkeys or gibbons presented with mirrors, despite repeated trials by Gallup and others. When the researchers dyed portions of the faces of chimpanzees under anesthesia, the apes subsequently gave even more convincing evidence that they were self-aware. They spent more time at the mirrors, intently examining the changes in their appearance and smelling the fingers with which they had touched the altered areas.

If consciousness of self and the ability to communicate ideas with other intelligent beings exist, can other qualities of the human mind be far away? Premack has pondered the implications of transmitting the concept of personal death to chimpanzees, but he is hesitant. "What if, like man," he asks,

> the ape dreads death and will deal with this knowledge as bizarrely as we have? . . . The desired objective would be not only to communicate the knowledge of death but, more important, to find a way of making sure the apes' response would not be that of dread, which, in the human case, has led to the invention of ritual, myth, and religion. Until I can suggest concrete steps in teaching the concept of death without fear, I have no intention of imparting the knowledge of mortality to the ape.

And what of the social existence of the chimpanzees? They are far less elaborately organized than even the hunter-gatherers, who have the simplest economic arrangements of all human beings. Yet striking basic similarities exist. The apes live in troops of up to fifty individuals, within which smaller, more casual groups break off and reunite in shifting combinations of individuals over periods as brief as a few days. Males are somewhat larger than females, to about the same degree as in human beings, and they occupy the top of well-marked dominance hierarchies. Children are closely associated with their mothers over a period of years, sometimes even into maturity. The young chimpanzees themselves remain allied for long periods of time; individuals on occasion even adopt younger brothers or sisters when the mother dies.

Each troop occupies a home range of about twenty square miles. Meetings between neighboring troops are infrequent and usually tense. On these occasions nubile females and young mothers sometimes migrate between the groups. But on other occasions chimpanzees can become territorial and murderous. At the Gombe Stream Reserve in Tanzania, where Jane Goodall conducted her celebrated research, bands of males from one troop, encroaching on the home range of an adjacent, smaller troop, attacked and occasionally injured the defenders. Eventually the residents abandoned their land to the invaders.

Like primitive human beings, chimpanzees gather fruit and other vegetable foods primarily and hunt only secondarily. The difference between their diets is one of proportion. Where all of hunter-gatherer societies considered together derive an average of 35 percent of their calories from fresh meat, chimpanzees obtain between 1 and 5 percent. And whereas primitive human hunters capture prey of any size, including elephants one hundred times the weight of a man, chimpanzees rarely attack any animal greater than one-fifth the weight of an adult male. Perhaps the most remarkable form of man-like behavior among chimpanzees is the use of intelligent, cooperative maneuvers during the hunt. Normally only adult males attempt to pursue animals — another humanoid trait. When a potential victim, such as a vervet or young baboon, has been selected, the chimpanzees signal their intentions by distinctive changes in posture, movement, and facial expression. Other males respond by turning to stare at the target animal. Their posture is tensed, their hair partially erected, and they become silent — a conspicuous change from the human observer's point of view, because chimpanzees are ordinarily the noisiest of animals. The state of alertness is broken by a sudden, nearly simultaneous pursuit.

A common strategy of the hunter males is to mingle with a group of baboons and then attempt to seize one of the youngsters with an explosive rush. Another is to encircle and stalk the victim, even while it nervously edges away. At the Gombe Stream Reserve an enterprising male named Figan tracked a juvenile baboon until it retreated up the trunk of a palm tree. Within moments other males that had been resting and grooming nearby stood up and walked over to join the pursuit. A few stopped at the bottom of the tree in which the baboon waited, while others dispersed to the bases of adjacent trees that might have served as alternate routes of escape. The baboon then leaped onto a second tree, whereupon the chimpanzee stationed below began to climb quickly toward it. The baboon finally managed to escape by jumping twenty feet to the ground and running to the protection of its troop nearby.

The distribution of the meat is also cooperative, with favors asked and given. The begging chimpanzee stares intently while holding its face close to the meat or to the face of the meat eater. It may also reach out and touch the meat and the chin and lips of the other animal, or extend an open hand with palm upward beneath his chin. Sometimes the male holding the prey moves abruptly away. But often he acquiesces by allowing the other animal to chew directly on the meat or to remove small pieces with its hands. On a few occasions males go so far as to tear off pieces of meat and hand them over to supplicants. This is a small gesture by the standards of human altruism but it is a very rare act among animals — a giant step, one might say, for apekind.

Finally, chimpanzees have a rudimentary culture. During twenty-five years of research on free-living troops in the forests of Africa, teams of zoologists from Europe, Japan, and the United States have discovered a remarkable repertory of tool use in the ordinary life of the apes. It includes the use of sticks and saplings as defensive weapons against leopards; the hurling of sticks, stones, and handfuls of vegetation during attacks on baboons, human beings, and other chimpanzees; digging with sticks to tear open termite mounds and "fishing" for the termites with plant stems stripped of leaves and split down the middle; prying open boxes with sticks; and lifting water from tree holes in "sponges" constructed of chewed leaves.

Learning and play are vital to the acquisition of the tool-using skills. When two-year-old chimpanzee infants are denied the opportunity to play with sticks their ability to solve problems with the aid of sticks at a later age is reduced. Given access to play objects, young animals in captivity progress through a relatively invariant maturation of skills. Under two years of age they simply touch or hold objects without attempting to manipulate them. As they grow older they increasingly employ one object to hit or prod another, while simultaneously improving in the solution of problems that require the use of tools. A similar progression occurs in the wild populations of Africa. Infants as young as six weeks reach out from their mother's clasp to fondle leaves and branches. Older infants constantly inspect their environment with their eyes, lips, tongues, noses, and hands, while periodically plucking leaves and waving them about. During this development they advance to tool-using behavior in small steps. One eight-month-old infant was seen to add grass stems to his other toys — but for the special purpose of wiping them against other objects, such as stones and his mother. This is the behavior pattern uniquely associated with termite "fishing" — by which the apes provoke the insects into running onto the object and then quickly bite or lick them off. During play, other infants prepared grass stalks as fishing tools by shredding the edges off wide blades and chewing the ends off long stems.

Jane Goodall has obtained direct evidence of imitative behavior in the transmission of these traditions. She observed infants watch adults as they used tools, then pick the tools up and use them after the adults had moved away. On two occasions a three-year-old youngster was seen to observe his mother closely as she wiped dung from her bottom with leaves. Then he picked up leaves and imitated the movements, even though his bottom was not dirty.

Chimpanzees are able to invent techniques and to transmit them to others. The use of sticks to pry open food boxes is a case in point. The method was invented by one or a few individuals at the Gombe Stream Reserve, then evidently spread through the troop by imitation. One female new to the area remained hidden in the bushes while watching others trying to open the boxes. On her fourth visit she walked into the open, picked up a stick, and began to poke it at the boxes.

Each tool-using behavior recorded in Africa is limited to certain populations of chimpanzees but has a mostly continuous distribution within its range. This is just the pattern expected if the behavior had been spread culturally. Maps of chimpanzee tool-using recently prepared by the Spanish zoologist Jorge Sabater-Pí might be placed without notice into a chapter on primitive culture in an anthropology textbook. Although most of the evidence concerning invention and transmission of the tool-using methods is indirect, it suggests that the apes have managed to cross the threshold of cultural evolution and thus, in an important sense, to have moved on into the human domain.

This account of the life of the chimpanzee is meant to establish what I regard as a fundamental point about the human condition: that by conventional evolutionary measures and the principal criteria of psychology we are not alone, we have a little-brother species. The points of similarity between human and chimpanzee social behavior, when joined with the compelling anatomical and biochemical traces of relatively recent genetic divergence, form a body of evidence too strong to be dismissed as coincidence. I now believe that they are based at least in part on the possession of identical genes. If this proposition contains any truth, it makes even more urgent the conservation and closer future study of these and the other great apes, as well as the Old World monkeys and the lower primates. A more thorough knowledge of these animal species might well provide us with a clearer picture of the step-by-step genetic changes that led to the level of evolution uniquely occupied by human beings.

To summarize the argument to this point: the general traits of human nature appear limited and idiosyncratic when placed against the great backdrop of all other living species. Additional evidence suggests that the more stereotyped forms of human behavior are mammalian and even more specifically primate in character, as predicted on the basis of general evolutionary theory. Chimpanzees are close enough to ourselves in the details of their social life and mental properties to rank as nearly human in certain domains where it was once considered inappropriate to make comparisons at all. These facts are in accord with the hypothesis that human social behavior rests on a genetic foundation — that human behavior is, to be more precise, organized by some genes that are shared with closely related species and others that are unique to the human species. The same facts are unfavorable for the competing hypothesis which has dominated the social sciences for generations, that mankind has escaped its own genes to the extent of being entirely culture-bound.

Let us pursue this matter systematically. The heart of the genetic hypothesis is the proposition, derived in a straight line from neo-Darwinian evolutionary theory, that the traits of human nature were adaptive during the time that the human species evolved and that genes consequently spread through the population that predisposed their carriers to develop those traits. Adaptiveness means simply that if an individual displayed the traits he stood a greater chance of having his genes represented in the next generation than if he did not display the traits. The differential advantage among individuals in this strictest sense is called genetic fitness. There are three basic components of genetic fitness: increased personal survival, increased personal reproduction, and the enhanced survival and reproduction of close relatives who share the same genes by common descent. An improvement in any one of the factors or in any combination of them results in greater genetic fitness. The process, which Darwin called natural selection, describes a tight circle of causation. If the possession of certain genes predisposes individuals toward a particular trait, say a certain kind of social response, and the trait in turn conveys superior fitness, the genes will gain an increased representation in the next generation. If natural selection is continued over many generations, the favored genes will spread throughout the population, and the trait will become characteristic of the species. In this way human nature is postulated by many sociobiologists, anthropologists, and others to have been shaped by natural selection.

It is nevertheless a curious fact, which enlarges the difficulty of the analysis, that sociobiological theory can be obeyed by purely cultural behavior as well as by genetically constrained behavior. An almost purely cultural sociobiology is possible. If human beings were endowed with nothing but the most elementary drives to survive and to reproduce, together with a capacity for culture, they would still learn many forms of social behavior that increase their biological fitness. But as I will show, there is a limit to the amount of this cultural mimicry, and methods exist by which it can be distinguished from the more structured forms of biological adaptation. The analysis will require the careful use of techniques in biology, anthropology, and psychology. Our focus will be on the closeness of fit of human social behavior to sociobiological theory, and on the evidences of genetic constraint seen in the strength and automatic nature of the predispositions human beings display while developing this behavior.

Let me now rephrase the central proposition in a somewhat stronger and more interesting form: if the genetic components of human nature did not originate by natural selection, fundamental evolutionary theory is in trouble. At the very least the theory of evolution would have to be altered to account for a new and as yet unimagined form of genetic change in populations. Consequently, an auxiliary goal of human sociobiology is to learn whether the evolution of human nature conforms to conventional evolutionary theory. The possibility that the effort will fail conveys to more adventurous biologists a not unpleasant whiff of grapeshot, a crackle of thin ice.

We can be fairly certain that most of the genetic evolution of human social behavior occurred over the five million years prior to civilization, when the species consisted of sparse, relatively immobile populations of hunter-gatherers. On the other hand, by far the greater part of cultural evolution has occurred since the origin of agriculture and cities approximately 10,000 years ago. Although genetic evolution of some kind continued during this latter, historical sprint, it cannot have fashioned more than a tiny fraction of the traits of human nature. Otherwise surviving hunter-gatherer people would differ genetically to a significant degree from people in advanced industrial nations, but this is demonstrably not the case. It follows that human sociobiology can be most directly tested in studies of hunter-gatherer societies and the most persistent preliterate herding and agricultural societies. As a result, anthropology rather than sociology or economics is the social science closest to sociobiology. It is in anthropology that the genetic theory of human nature can be most directly pursued.

The power of a scientific theory is measured by its ability to transform a small number of axiomatic ideas into detailed predictions of observable phenomena; thus the Bohr atom made modern chemistry possible, and modern chemistry recreated cell biology. Further, the validity of a theory is measured by the extent to which its predictions successfully compete with other theories in accounting for the phenomena; the solar system of Copernicus won over that of Ptolemy, after a brief struggle. Finally, a theory waxes in influence and esteem among scientists as it assembles an ever larger body of facts into readily remembered and usable explanatory schemes, and as newly discovered facts conform to its demands: the round earth is more plausible than a flat one. Facts crucial to the advancement of science can be obtained either by experiments designed for the purpose of acquiring them or from the inspired observation of undisturbed natural phenomena. Science has always progressed in approximately this opportunistic, zig-zagging manner.

In the case of the theory of the genetic evolution of human nature, if it is ever to be made part of real science, we should be able to select some of the best principles from ecology and genetics, which are themselves based on the theory, and adapt them in detail to human social organization. The theory must not only account for many of the known facts in a more convincing manner than traditional explanations, but must also identify the need for new kinds of information previously unimagined by the social sciences. The behavior thus explained should be the most general and least rational of the human repertoire, the part furthest removed from the influence of day-to-day reflection and the distracting vicissitudes of culture. In other words, they should implicate innate, biological phenomena that are the least susceptible to mimicry by culture.

These are stern requirements to impose on the infant discipline of human sociobiology, but they can be adequately justified. Sociobiology intrudes into the social sciences with credentials from the natural sciences and, initially, an unfair psychological advantage. If the ideas and analytical methods of "hard" science can be made to work in a congenial and enduring manner, the division between the two cultures of science and the humanities will close. But if our conception of human nature is to be altered, it must be by means of truths conforming to the canons of scientific evidence and not a new dogma however devoutly wished for.

Various sociobiological explorations in the deeper mode, some already reasonably secure and others frankly speculative, are the theme of the next six chapters of this book. For the moment, to illustrate the method, let me present two concise examples.

Incest taboos are among the universals of human social behavior. The avoidance of sexual intercourse between brothers and sisters and between parents and their offspring is everywhere achieved by cultural sanctions. But at least in the case of the brother-sister taboo, there exists a far deeper, less rational form of enforcement: a sexual aversion automatically develops between persons who have lived together when one or all grew to the age of six. Studies in Israeli kibbutzim, the most thorough of which was conducted by Joseph Shepher of the University of Haifa, have shown that the aversion among people of the same age is not dependent on an actual blood relationship. Among 2,769 marriages recorded, none was between members of the same kibbutz peer group who had been together since birth. There was not even a single recorded instance of heterosexual activity, despite the fact that the kibbutzim adults were not opposed to it. Where incest of any form does occur at low frequencies in less closed societies, it is ordinarily a source of shame and recrimination. In general, mother-son intercourse is the most offensive, brother-sister intercourse somewhat less and father-daughter intercourse the least offensive. But all forms are usually proscribed. In the United States at the present time, one of the forms of pornography considered most shocking is the depiction of intercourse between fathers and their immature daughters.

What advantage do the incest taboos confer? A favored explanation among anthropologists is that the taboos preserve the integrity of the family by avoiding the confusion in roles that would result from incestuous sex. Another, originated by Edward Tylor and built into a whole anthropological theory by Claude Lévi-Strauss in his seminal *Les Structures Élémentaires de la Parenté*, is that it facilitates the exchange of women during bargaining between social groups. Sisters and daughters, in this view, are not used for mating but to gain power.

In contrast, the prevailing sociobiological explanation regards family integration and bridal bargaining as by-products or at most as secondary contributing factors. It identifies a deeper, more urgent cause, the heavy physiological penalty imposed by inbreeding. Several studies by human geneticists have demonstrated that even a moderate amount of inbreeding results in children who are diminished in overall body size, muscular coordination, and academic performance. More than one hundred recessive genes have been discovered that cause hereditary disease in the undiluted, homozygous state, a condition vastly enhanced by inbreeding. One analysis of American and French populations produced the estimate that each person carries an average of four lethal gene equivalents: either four genes that cause death outright when in the homozygous state, eight genes that cause death in fifty percent of homozygotes, or other, arithmetically equivalent combinations of lethal and debilitating effects. These high numbers, which are typical of animal species, mean that inbreeding carries a deadly risk. Among 161 children born to Czechoslovakian women who had sexual relations with their fathers, brothers, or sons, fifteen were stillborn or died within the first year of life, and more than 40 percent suffered from various physical and mental defects, including severe mental retardation, dwarfism, heart and brain deformities, deaf-mutism, enlargement of the colon, and urinary-tract abnormalities. In contrast, a group of ninety-five children born to the same women through nonincestuous relations were on the average as normal as the population at large. Five died during the first year of life, none had serious mental deficiencies, and only five others had apparent physical abnormalities.

The manifestations of inbreeding pathology constitute natural selection in an intense and unambiguous form. The elementary theory of population genetics predicts that any behavioral tendency to avoid incest, however slight or devious, would long ago have spread through human populations. So powerful is the advantage of outbreeding that it can be expected to have carried cultural evolution along with it. Family integrity and leverage during political bargaining may indeed be felicitous results of outbreeding, but they are more likely to be devices of convenience, secondary cultural adaptations that made use of the inevitability of outbreeding for direct biological reasons.

Of the thousands of societies that have existed through human history, only several of the most recent have possessed any knowledge of genetics. Very few opportunities presented themselves to make rational calculations of the destructive effects of inbreeding. Tribal councils do not compute gene frequencies and mutational loads. The automatic exclusion of sexual bonding between individuals who have previously formed certain other kinds of relationships — the "gut feeling" that promotes the ritual sanctions against incest — is largely unconscious and irrational. Bond exclusion of the kind displayed by the Israeli children is an example of what biologists call a proximate (near) cause; in this instance, the direct psychological exclusion is the proximate cause of the incest taboo. The ultimate cause suggested by the biological hypothesis is the loss of genetic fitness that results from incest. It is a fact that incestuously produced children leave fewer descendants. The biological hypothesis states that individuals with a genetic predisposition for bond exclusion and incest avoidance contribute more genes to the next generation. Natural selection has probably ground away along these lines for thousands of generations, and for that reason human beings intuitively avoid incest through the simple, automatic rule of bond exclusion. To put the idea in its starkest form, one that acknowledges but temporarily bypasses the intervening developmental process, human beings are guided by an instinct based on genes.

Such a process is indicated in the case of brother-sister intercourse, and it is a strong possibility in the other categories of incest taboo.

Hypergamy is the female practice of marrying men of equal or greater wealth and status. In human beings and most kinds of social animals, it is the females who move upward through their choice of mates. Why this sexual bias? The vital clue has been provided by Robert L. Trivers and Daniel E. Willard in the course of more general work in sociobiology. They noted that in vertebrate animals generally, and especially birds and mammals, large, healthy males mate at a relatively high frequency while many smaller, weaker males do not mate at all. Yet nearly all females mate successfully. It is further true that females in the best physical condition produce the healthiest infants, and these offspring usually grow up to be the largest, most vigorous adults. Trivers and Willard then observed that according to the theory of natural selection females should be expected to give birth to a higher proportion of males when they are healthiest, because these offspring will be largest in size, mate most successfully, and produce the maximum number of offspring. As the condition of the females deteriorates, they should shift progressively to the production of daughters, since female offspring will now represent the safer investment. According to natural-selection theory, genes that induce this reproductive strategy will spread through the population at the expense of genes that promote alternative strategies.

It works. In deer and human beings, two of the species investigated with reference to this particular question, environmental conditions adverse for pregnant females are associated with a disproportionate increase in the birth of daughters. Data from mink, pigs, sheep, and seals also appear to be consistent with the Trivers-Willard prediction. The most likely direct mechanism is the selectively greater mortality of male fetuses under adversity, a phenomenon that has been documented in numerous species of mammals.

Altering the sex ratio before birth is of course an entirely irrational act; it is in fact physiological. Mildred Dickeman, an anthropologist, has tested the theory in the realm of conscious behavior. She has asked whether the sex ratio is altered after birth by infanticide in a way that fits the best reproductive strategy. Such appears to be the case. In precolonial and British India, the upward social flow of daughters by marriage to higher ranking men was sanctified by rigid custom and religion, while female infanticide was practiced routinely by the upper castes. The Bedi Sikhs, the highest ranking priestly subcaste of the Punjab, were known as *Kuri-Mar*, the daughter slayers. They destroyed virtually all female infants and invested everything in raising sons who would marry women from lower castes. In pre-revolutionary China, female infanticide was commonly practiced by many of the social classes, with essentially the same effects as in India — that is, a socially upward flow of women accompanied by dowries, a concentration of both wealth and women in the hands of a small middle and upper class, and near exclusion of the poorest males from the breeding system. It remains to be seen whether this pattern is widespread in human cultures. For the moment the existence of even a few cases suggests the need for a reexamination of the phenomenon with close attention to biological theory.

Female hypergamy and infanticide do not recommend themselves as rational processes. It is difficult to explain them except as an inherited predisposition to maximize the number of offspring in competition with other members of the society. Research of the kind started by Dickeman, if extended to other societies, will help to test this proposition more rigorously. If successful, it can be expected to shed light on the deeper mental processes that move people to choose one complicated course of action out of the many open, in principle, to rational choice.

Human nature can be probed by other, more directly psychological techniques. Behavior that is both irrational and universal should also be more resistant to the distorting effects of cultural deprivation than more intellectual, individualistic behavior, and less likely to be influenced by the frontal lobes and the other higher centers of the brain that serve as the headquarters of long-term rational thought. Such behavior is more likely to be heavily influenced by the limbic system, the evolutionarily ancient portion of the cortex located near the physical center of the brain. Given that the higher and lower controls in the brain are anatomically separated to some extent, we can expect to find occasional human beings whose rational faculties have been impaired for one reason or another but who continue to function well at the level of instinct.

Such persons exist. In his study of patients in institutions for the mentally retarded, Richard H. Wills has found that two distinct types can be identified. "Cultural retardates" have well below normal intelligence, but their behavior retains many uniquely human attributes. They communicate with attendants and one another by speech, and they initiate a variety of relatively sophisticated actions, such as singing alone and in groups, listening to records, looking at magazines, working at simple tasks, bathing, grooming themselves, smoking cigarettes, exchanging clothing, teasing and directing others, and volunteering favors. The second group, the "noncultural retardates," represent a sudden and dramatic step downward in ability. They perform none of the actions just listed. Their exchanges with others entail little that can be labeled as truly human communication. Cultural behavior thus seems to be a psychological whole invested in the brain or denied it in a single giant step. Yet the noncultural retardates retain a large repertory of more "instinctive" behavior, the individual actions of which are complex and recognizably mammalian. They communicate with facial expressions and emotion-laden sounds, examine and manipulate objects, masturbate manually, watch others, steal, stake out small territories, defend themselves, and play, both as individuals and in groups. They frequently seek physical contact with others; they offer and solicit affection by means of strongly expressed, unmistakable gestures. Virtually none of their responses is abnormal in a biological sense. Fate has merely denied these patients entry into the cultural world of the brain's outer cortex.

Let me now try to answer the important but delicate question of how much social behavior varies genetically *within* the human species. The fact that human behavior still has structure based on physiology and is mammalian in its closest affinities suggests that it has been subject to genetic evolution until recently. If that is true, genetic variation affecting behavior might even have persisted into the era of civilization. But this is not to say that such variation now exists.

Two possibilities are equally conceivable. The first is that in reaching its present state the human species exhausted its genetic variability. One set of human genes affecting social behavior, and one set only, survived the long trek through prehistory. This is the view implicitly favored by many social scientists and, within the spectrum of political ideologies that address such questions, by many intellectuals of the left. Human beings once evolved, they concede, but only to the point of becoming a uniform, language-speaking, culture-bearing species. By historical times mankind had become magnificent clay in the hands of the environment. Only cultural evolution can now occur. The second possibility is that at least some genetic variation still exists. Mankind might have ceased evolving, in the sense that the old biological mode of natural selection has relaxed its grip, but the species remains capable of both genetic and cultural evolution.

The reader should note that either possibility — complete cultural determination versus shared cultural and genetic determination of variability within the species — is compatible with the more general sociobiological view of human nature, namely that the most diagnostic features of human behavior evolved by natural selection and are today constrained throughout the species by particular sets of genes.

These possibilities having been laid out in such a textbook fashion, I must now add that the evidence is strong that a substantial fraction of human behavioral variation is based on genetic differences among individuals. There are undeniably mutations affecting behavior. Of these changes in the chemical composition of genes or the structure and arrangement of chromosomes, more than thirty have been identified that affect behavior, some by neurological disorders, others by the impairment of intelligence. One of the most controversial but informative examples is the XYY male. The X and Y chromosomes determine sex in human beings; the XX combination produces a female, XY a male. Approximately 0.1 percent of the population accidentally acquires an extra Y chromosome at the moment of conception, and these XYY individuals are all males. The XYY males grow up to be tall men, the great majority over six feet. They also end up more frequently in prisons and hospitals for the criminally insane. At first it was thought that the extra chromosome induced more aggressive behavior, creating what is in effect a class of genetic criminals. However, a statistical study, by Princeton psychologist Herman A. Witkin and his associates, of vast amounts of data from Denmark has led to a more benign interpretation. XYY men were found neither to be more aggressive than normal nor to display any particular behavior pattern distinguishing them from the remainder of the Danish population. The only deviation detected was a lower average intelligence. The most parsimonious explanation is that XYY men are incarcerated at a higher rate because they are simply less adroit at escaping detection. However, caution is required. The possibility of the inheritance of more specific forms of predisposition toward a criminal personality has not been excluded by this one study.

In fact, mutations have been identified that do alter specific features of behavior. Turner's syndrome, occurring when only one of the two X chromosomes is passed on, entails not just a lowered general intelligence but a particularly deep impairment in the ability to recall shapes and to orient between the left and right on maps and other diagrams. The Lesch-Nyhan syndrome, induced by a single recessive gene, causes both lowered intelligence and a compulsive tendency to pull and tear at the body, resulting in self-mutilation. The victims of these and other genetic disorders, like the severely mentally retarded, provide extraordinary opportunities for a better understanding of human behavior. The form of analysis by which they can be most profitably studied is called genetic dissection. Once a condition appears, despite medical precautions, it can be examined closely in an attempt to pinpoint the altered portion of the brain and to implicate hormones and other chemical agents that mediated the change without, however, physically touching the brain. Thus by the malfunctioning of its parts the machine can be diagrammed. And let us not fall into the sentimentalist trap of calling that procedure cold-blooded; it is the surest way to find a medical cure for the conditions themselves.

Most mutations strong enough to be analyzed as easily as the Turner and Lesch-Nyhan anomalies also cause defects and illnesses. This is as true in animals and plants as it is in human beings, and is entirely to be expected. To understand why, consider the analogy of heredity with the delicate construction of a watch. If a watch is altered by randomly shaking or striking it, as the body's chemistry is randomly transformed by a mutation, the action is far more likely to impair than to improve the accuracy of the watch.

This set of strong examples, however, leaves unanswered the

question of the genetic variation and evolution of "normal" social behavior. As a rule, traits as complex as human behavior are influenced by many genes, each of which shares only a small fraction of the total control. These "polygenes" cannot ordinarily be identified by detecting and tracing the mutations that alter them. They must be evaluated indirectly by statistical means. The most widely used method in the genetics of human behavior is the comparison of pairs of identical twins with pairs of fraternal twins. Identical twins originate in the womb from a single fertilized ovum. The two cells produced by the first division of the ovum do not stick together to produce the beginnings of the fetus but instead separate to produce the beginnings of two fetuses. Because the twins originated from the same cell, bearing a single nucleus and set of chromosomes, they are genetically identical. Fraternal twins, in contrast, originate from separate ova that just happen to travel into the reproductive tracts and to be fertilized by different sperm at the same time. They produce fetuses genetically no closer to one another than are brothers or sisters born in different years.

Identical and fraternal twins provide us with a natural controlled experiment. The control is the set of pairs of identical twins: any differences between the members of a pair must be due to the environment (barring the very rare occurrence of a brand-new mutation). Differences between the members of a pair of fraternal twins can be due to their heredity, their environment, or to some interaction between their heredity and environment. If in a given trait, such as height or nose shape, identical twins prove to be closer to one another on the average than are fraternal twins of the same sex, the difference between the two kinds of twins can be taken as prima facie evidence that the trait is influenced to some degree by heredity. Using this method, geneticists have implicated heredity in the formation of a variety of traits that affect social relationships: number ability, word fluency, memory, the timing of language acquisition, spelling, sentence construction, perceptual skill, psycho-motor skill, extroversion-introversion, homosexuality, the age of first sexual activity, and certain forms of neurosis and psychosis, including manic-depressive behavior and schizophrenia.

There is a catch in these results that render them less than defini-tive. Identical twins are regularly treated alike by their parents, more so than fraternal twins. They are more frequently dressed alike, kept together for longer times, fed the same way, and so on. Thus in the absence of other information it is possible that the greater sim-ilarity of identical twins could, after all, be due to the environment. However, there exist new, more sophisticated techniques that can take account of this additional factor. Such a refinement was em-ployed by the psychologists John C. Loehlin and Robert C. Nichols in their analysis of the backgrounds and performances of 850 sets of twins who took the National Merit Scholarship test in 1962. Not only the differences between identical and fraternal twins, but also the early environments of all the subjects were carefully examined and weighed. The results showed that the generally closer treat-ment of identical twins is not enough to account for their greater similarity in general abilities, personality traits, or even ideals, goals, and vocational interests. The conclusion to be drawn is that either the similarities are based in substantial part on genetic closeness, or else environmental factors were at work that remained hidden to the psychologists.

My overall impression of the existing information is that *Homo sapiens* is a conventional animal species with reference to the quality and magnitude of the genetic diversity affecting its behavior. If the comparison is correct, the psychic unity of mankind has been reduced in status from a dogma to a testable hypothesis.

I also believe that it will soon be within our power to identify many of the genes that influence behavior. Thanks largely to ad-vances in techniques that identify minute differences in the chemical products prescribed by genes, our knowledge of the fine details of human heredity has grown steeply during the past twenty years. In 1977 the geneticists Victor McKusick and Francis Ruddle reported in *Science* that twelve hundred genes had been distinguished; of these, the position of 210 had been pinpointed to a particular chromosome, and at least one gene had been located on each of the twenty-three pairs of chromosomes. Most of the genes ultimately affect anatomical and biochemical traits having minimal influence on behavior. Yet some do affect behavior in important ways, and a few of the behavioral mutations have been closely linked to known biochemical changes. Also, subtle behavioral controls are known that incorporate alterations in levels of hormones and transmitter substances acting directly on nerve cells. The recently discovered enkephalins and endorphins are protein-like substances of relatively simple structure that can profoundly affect mood and temperament. A single mutation altering the chemical nature of one or more of them might change the personality of the person bearing it, or at least the predisposition of the person to develop one personality as opposed to another in a given cultural surrounding. Thus it is possible, and in my judgment even probable, that the positions of genes having indirect effects on the most complex forms of behavior will soon be mapped on the human chromosomes. These genes are unlikely to prescribe particular patterns of behavior; there will be no mutations for a particular sexual practice or mode of dress. The behavioral genes more probably influence the ranges of the form and intensity of emotional responses, the thresholds of arousals, the readiness to learn certain stimuli as opposed to others, and the pattern of sensitivity to additional environmental factors that point cultural evolution in one direction as opposed to another.

It is of equal interest to know whether even "racial" differences in behavior occur. But first I must issue a strong caveat, because this is the most emotionally explosive and politically dangerous of all subjects. Most biologists and anthropologists use the expression "racial" only loosely, and they mean to imply nothing more than the observation that certain traits, such as average height or skin color, vary genetically from one locality to another. If Asians and Europeans are said to differ from one another in a given property, the statement means that the trait changes in some pattern between Asia and Europe. It does not imply that discrete "races" can be defined on the basis of the trait, and it leaves open a strong possibility that the trait shows additional variation within different parts of Asia and Europe. Furthermore, various properties in anatomy and physiology — for example, skin color and the ability to digest milk — display widely differing patterns of geographical ("racial") variation. As a consequence most scientists have long recognized that it is a futile exercise to try to define discrete human races. Such entities do not in fact exist. Of equal importance, the description of geographical variation in one trait or another by a biologist or anthropologist or anyone else should not carry with it value judgments concerning the worth of the characteristics defined.

Now we are prepared to ask in a more fully objective manner: Does geographical variation occur in the genetic basis of social behavior? The evidence is strong that almost all differences between human societies are based on learning and social conditioning rather than on heredity. And yet perhaps not quite all. Daniel G. Freedman, a psychologist at the University of Chicago, has addressed this question with a series of studies on the behavior of newborn infants of several racial origins. He has detected significant average differences in locomotion, posture, muscular tone of various parts of the body, and emotional response that cannot reasonably be explained as the result of training or even conditioning within the womb. Chinese-American newborns, for example, tend to be less changeable, less easily perturbed by noise and movement, better able to adjust to new stimuli and discomfort, and quicker to calm themselves than

Caucasian-American infants. To use a more precise phrasing, it can be said that a random sample of infants whose ancestors originated in certain parts of China differ in these behavioral traits from a comparable sample of European ancestry.

There is also some indication that the average differences carry over into childhood. One of Freedman's students, Nova Green, found that Chinese-American children in Chicago nursery schools spent less of their time in approach and interaction with playmates and more time on individual projects than did their European-American counterparts. They also displayed interesting differences in temperament:

> Although the majority of the Chinese-American children were in the "high arousal age," between 3 and 5, they showed little intense emotional behavior. They ran and hopped, laughed and called to one another, rode bikes and roller-skated just as the children did in the other nursery schools, but the noise level stayed remarkably low and the emotional atmosphere projected serenity instead of bedlam. The impassive facial expression certainly gave the children an air of dignity and self-possession, but this was only one element affecting the total impression. Physical movements seemed more coordinated, no tripping, falling, bumping or bruising was observed, no screams, crashes or wailing was heard, not even that common sound in other nurseries, voices raised in highly indignant moralistic dispute! No property disputes were observed and only the mildest version of "fighting behavior," some good natured wrestling among the older boys.

Navaho infants tested by Freedman and his coworkers were even more quiescent than the Chinese infants. When lifted erect and pulled forward they were less inclined to swing their legs in a walking motion; when put in a sitting position, their backs curved;

and when placed on their stomachs, they made fewer attempts to crawl. It has been conventional to ascribe the passivity of Navaho children to the practice of cradleboarding, a device that holds the infant tightly in place on the mother's back. But Freedman suggests that the reverse may actually be true: the relative quiescence of Navaho babies, a trait that is apparent from birth onward, allows them to be carried in a confining manner. Cradleboarding represents a workable compromise between cultural invention and infant constitution.

Given that humankind is a biological species, it should come as no shock to find that populations are to some extent genetically diverse in the physical and mental properties underlying social behavior. A discovery of this nature does not vitiate the ideals of Western civilization. We are not compelled to believe in biological uniformity in order to affirm human freedom and dignity. The sociologist Marvin Bressler has expressed this idea with precision: "An ideology that tacitly appeals to biological equality as a condition for human emancipation corrupts the idea of freedom. Moreover, it encourages decent men to tremble at the prospect of 'inconvenient' findings that may emerge in future scientific research. This unseemly anti-intellectualism is doubly degrading because it is probably unnecessary."

I will go further and suggest that hope and pride and not despair are the ultimate legacy of genetic diversity, because we are a single species, not two or more, one great breeding system through which genes flow and mix in each generation. Because of that flux, mankind viewed over many generations shares a single human nature within which relatively minor hereditary influences recycle through ever changing patterns, between the sexes and across families and entire populations. To understand the enormous significance of this biological unity, imagine our moral distress if australopithecine man-apes had survived to the present time, halfway in intelligence be-

Heredity

tween chimpanzees and human beings, forever genetically separated from both, evolving just behind us in language and the higher faculties of reason. What would be our obligation to them? What would the theologians say — or the Marxists, who might see in them the ultimate form of an oppressed class? Should we divide the world, guide their mental evolution to the human level, and establish a two-species dominion based on a treaty of intellectual and technological parity? Should we make certain they rose no higher? But even worse, imagine our predicament if we coexisted with a mentally superior human species, say *Homo superbus*, who regarded us, the minor sibling species *Homo sapiens*, as the moral problem.

Chapter 3. Development

The newly fertilized egg, a corpuscle one two-hundredth of an inch in diameter, is not a human being. It is a set of instructions sent floating into the cavity of the womb. Enfolded within its spherical nucleus are an estimated 250 thousand or more pairs of genes, of which fifty thousand will direct the assembly of the proteins and the remainder will regulate their rates of development. After the egg penetrates the blood-engorged wall of the uterus, it divides again and again. The expanding masses of daughter cells fold and crease into ridges, loops, and layers. Then, shifting like some magical kaleidoscope, they self-assemble into the fetus, a precise configuration of blood vessels, nerves, and other complex tissues. Each division and migration of the cells is orchestrated by a flow of chemical information that proceeds from the genes to the outer array of proteins, fats, and carbohydrates that make up the substance of the constituent cells.

In nine months a human being has been created. Functionally it is a digestive tube surrounded by sheaths of muscle and a skin. Its parts are continuously freshened with blood forced through closed blood vessels by the rhythmic pumping of the recently formed heart. The limited bodily actions are coordinated by an intricate interplay of hormones and nerves. The reproductive organs lie dormant; they await the precise hormonal signals that years later will trigger the second, final phase of their growth and call upon them to complete the organism's ultimate biological role. Atop this ensemble sits the brain. Its weight is one pound, its consistency that of thick custard, and its fine structure the most complicated machinery ever produced on earth. The brain contains an exact configuration of about ten billion neurons, or cellular units, each of which makes hundreds or even thousands of contacts with other neurons. Vast numbers of nerve fibers pass down from the brain through the spinal cord, where they connect with still other nerves that relay information and instructions back and forth to the remaining organs of the body. The central nervous system, comprising the brain and spinal cord in tandem, receives electrical signals from no fewer than a billion sensory elements, from the visual rods of the retina to the pressure-sensitive corpuscles of the skin.

The newborn infant is now seen to be wired with awesome precision. The movements of its eyes are steered by thousands of nerve cells that fan out from the eye muscles to reflex stations between the eye and brain, as well as by higher integrating centers scattered over the frontal eyefields and other centers of the brain's cortex. The baby listens: sounds of each frequency activate a particular cluster of receptors in the inner ear, which pass signals to corresponding masses of nerve cells at successively higher levels of the brain. The signals proceed inward, as though melodies were being played on a piano keyboard projected from the inner ear, then again by a new diatonic scale at way-stations in the hindbrain, next at the inferior colliculi of the midbrain and the medial geniculate bodies of the forebrain, and finally at the auditory cortex of the forebrain, where in some manner beyond our present understanding the mind "hears" the sound.

This marvelous robot is launched into the world under the care of its parents. Its rapidly accumulating experience will soon transform it into an independently thinking and feeling individual. Then the essential components of social behavior will be added—language, pair bonding, rage at ego injury, love, tribalism, and all the remainder of the human-specific repertory. But to what extent does the wiring of the neurons, so undeniably encoded in the genes, preordain the directions that social development will follow? Is it possible that the wiring diagram has been constructed by evolution only to be an all-purpose device, adaptable through learning to any mode of social existence?

This then is the frame of reference by which we can grasp the full dimensions of the empirical problem of human behavior: from 250 thousand genes to ten billion neurons to an unknown potential variety of social systems. In the last chapter I used the comparison of mankind with species of social animals to demonstrate that contemporaneous human behavior is constrained by heredity. As anticipated by evolutionary theory, behavioral development is channeled in the direction of the most generally mammalian traits. But what is the ultimate range of our potential? How far can human beings be moved across or even outside the mammalian channels? The answer must be sought in the study of individual development with special reference to genetic determinism.

We have at last come to the key phrase: genetic determinism. On its interpretation depends the entire relation between biology and the social sciences. To those who wish to reject the implications of sociobiology out of hand, it means that development is insect-like, confined to a single channel, running from a given set of genes to the corresponding single predestined pattern of behavior. The life of a mosquito does fit this narrow conception perfectly. When a winged adult emerges from its pupal case, it has only a few days to complete a set of intricate maneuvers leading to the deposit of a set of fertilized eggs in organically contaminated water. Both sexes get swiftly to work. The whine created by the wingbeat of the female, so irritating to the human ear, is a love song to the male. With no previous experience he flies toward the sound. The whine of a female yellow-fever mosquito is between 450 and 600 hertz (cycles per second). In the laboratory, entomologists have attracted males simply by striking a tuning fork set at these frequencies. When a cheese cloth is placed over the tuning fork, some of the more excited mosquitoes attempt to mate with it. The female mosquito cannot afford to be quite so impetuous, yet the episodes of her life follow a rigid marching order prescribed by her genes. She seeks out human and other mammalian prey by their warmth or, in the case of some species, by the odor of lactic acid emanating from the skin. Alighting, she probes the skin with two microscopic, thread-like and sharpened stylets. The points are plunged through the skin in search of a blood vessel, much as oil prospectors sink a well. Sometimes they strike a vessel and sometimes not. The female of at least one species of mosquito identifies blood by the taste of a chemical called adenosine diphosphate (ADP) found in the red cells. The only apparent significance of ADP among the hundreds of available blood constituents is that it serves as an immediately accessible marker. Other, similarly arbitrary "sign stimuli" guide the mosquito to appropriate ponds and smaller bodies of water where she can lay her eggs in safety.

The mosquito is an automaton. It can afford to be nothing else. There are only about one hundred thousand nerve cells in its tiny head, and each one has to pull its weight. The only way to run accurately and successfully through a life cycle in a matter of days is by instinct, a sequence of rigid behaviors programmed by the genes to unfold swiftly and unerringly from birth to the final act of oviposition.

The channels of human mental development, in contrast, are circuitous and variable. Rather than specify a single trait, human genes prescribe the *capacity* to develop a certain array of traits. In some categories of behavior, the array is limited and the outcome can be altered only by strenuous training—if ever. In others, the array is vast and the outcome easily influenced.

An example of a restricted behavior is handedness. Each person is biologically predisposed to be either left- or right-handed. In present-day Western societies parents are relatively tolerant of the outcome in their children, who therefore follow the direction set by the genes affecting this trait. But traditional Chinese societies still exert a strong social pressure for right-handed writing and eating. In their recent study of Taiwanese children, Evelyn Lee Teng and her associates found a nearly complete conformity in these two activities but little or no effect on handedness in other activities not subjected to special training. Thus in this behavioral trait the genes have their way unless specifically contravened by conscious choice.

The evolution of capacity is illustrated in a still more graphic fashion by the genetic condition called phenylketonuria (PKU), which produces feeblemindedness as a physiological side effect. PKU is caused by the possession of a single pair of recessive genes among the hundreds of thousands of paired genes on the human chromosomes. Persons afflicted with a double dose of the PKU gene are unable to utilize a common dietary element, the amino acid phenylalanine. When the chemical breakdown of phenylalanine is blocked, abnormal intermediate products accumulate in the body. The urine turns dark on exposure to air and emits a distinctive mousy smell. One child out of approximately every ten thousand born has this genetic defect. Unless the poisoning is reversed by the time the PKU individual reaches the age of four to six months, he suffers an irreversible mental retardation. Fortunately, the disaster can be avoided by early diagnosis and restriction to a diet kept low in phenylalanine. In PKU the interaction between genes and environment is displayed in its simplest conceivable form. The infant born with two PKU genes has the capacity for either normal mental development or impairment, with a strong bias toward the latter. Only by making an extraordinary and very particular change in the environment—feeding the PKU infant a low-phenylalanine diet—can the bias be reversed. Thus, in order to predict with reasonable certainty whether any given newborn infant will have normal intelligence or succumb to the feeblemindedness of PKU, it is necessary to know both the genes and the environment.

Few behaviors are under the control of one or two genes, or can be turned on and off in the manner of PKU mental retardation. And even in the case of PKU, the trait is one of crude impairment rather than a subtle shift in patterns of response. A more typical relationship between genes and behavior is shown by schizophrenia, the commonest form of mental illness. Schizophrenia is not a simple cessation or distortion of normal behavior. A few psychiatrists, most notably Thomas Szasz and R. D. Laing, have viewed it as no more than an arbitrary label imposed by society on certain deviant individuals. But they have been proved almost certainly wrong. It is true that schizophrenia appears on the surface to be a purposeless mélange of odd responses. It consists of various combinations of hallucinations, delusions, inappropriate emotional responses, compulsively repeated movements of no particular significance, and even the deathlike immobilization of the catatonic trance. The variations are endlessly subtle, and psychiatrists have learned to treat each patient as a unique case. The borderline between normal and schizophrenic people is broad and nearly imperceptible. Mild schizophrenics function undetected among us in large numbers, while fully normal persons are sometimes erroneously diagnosed as schizophrenics. Nevertheless, three extreme kinds of schizophrenia are unmistakable: the haunted paranoid surrounded by his imaginary community of spies and assassins, the clownish, sometimes incontinent hebephrenic, and the frozen catatonic. Although the capacity to become schizophrenic may well be within all of us, there is no question that certain persons have distinctive genes predisposing them to the condition. Individuals taken from schizophrenic parents in infancy and placed with normal adoptive parents subsequently develop schizophrenic symptoms at a much higher rate than those given up for adoption by unafflicted parents. The data from hundreds of such cases have been analyzed painstakingly by Seymour Kety in collaboration with a team of American and Danish psychologists. Their results show conclusively that a major part of the tendency to become schizophrenic is inherited.

Evidence has also been adduced that schizophrenia is widespread in other kinds of human societies. Jane Murphy has found that both Eskimos from the Bering Sea and the Yorubas of Nigeria recognize and label a set of symptoms resembling the Western syndrome of schizophrenia. The afflicted individuals are, moreover, classified as mentally ill—their condition is called *nuthkavihak* by the Eskimos and *were* by the Yorubas—and they form a substantial fraction of the clientele of the tribal shamans and healers. The incidence of clear-cut schizophrenia is about the same as in Western societies; it ranges between 0.4 and 0.7 percent of the adult population.

Schizophrenia develops in a more complicated manner than PKU and most other hereditary forms of mental retardation. Whether a single gene or many genes are responsible is not known. Distinctive changes occur in the physiology of schizophrenics, and medical researchers may soon succeed in linking them directly to the mental aberrations. For example, Philip Seeman and Tyrone Lee have found that key areas of the brains of some schizophrenics contain twice the normal number of receptors for dopamine, a substance that carries signals between nerve cells. It is possible that this abnormality makes the brain unduly sensitive to its own signals and hence subject to hallucination. Yet the old psychological theories also have an element of truth: environment plays an important role in the development of the syndrome. There is such a thing as a typically "schizophrenogenic" (schizophrenia-producing) family arrangement, one most likely to produce a mentally ill adult from a child with the potential for the disease. In it trust has ended, communication has broken down, and the parents openly express contempt for each other while placing unreasonable demands on their children. Some psychiatrists even see a kind of twisted rationale in the mind of the schizophrenic: the individual tries to escape from his intolerable social environment by creating a private inner world. But the fact remains that certain genes predispose individuals toward schizophrenia. Individuals possessing them can develop the pathology while growing up in the midst of normal, supportive families.

Thus even in the relatively simple categories of behavior we inherit a *capacity* for certain traits, and a bias to learn one or another of those available. Scientists as diverse in their philosophies as Konrad Lorenz, Robert A. Hinde, and B. F. Skinner have often stressed that no sharp boundary exists between the inherited and the acquired. It has become apparent that we need new descriptive techniques to replace the archaic distinction between nature and nurture. One of the most promising is based on the imagery invented by Conrad H. Waddington, the great geneticist who died in 1975. Waddington said that development is something like a landscape that descends from highlands to the shore. Development of a trait—eye color, handedness, schizophrenia, or whatever—resembles the rolling of a ball down the slopes. Each trait traverses a different part of the landscape, each is guided by a different pattern of ridges and valleys. In the case of eye color, given a starting set of genes for blue or some other iris pigment, the topography is a single, deep channel. The ball rolls inexorably to one destination: once the egg has been joined by a sperm, only one eye color is possible. The developmental landscape of the mosquito can be similarly envisioned as a parallel series of deep, unbranching valleys, one leading to the sexual attraction of the wingbeat's sound, another to automatic bloodsucking, and so on through a repertory of ten or so discrete responses. The valleys form a precise, unyielding series of biochemical steps that proceed from the DNA in the fertilized egg to the neuromuscular actions mediated by the mosquito's brain.

The developmental topography of human behavior is enormously broader and more complicated, but it is still a topography. In some cases the valleys divide once or twice. An individual can end up either right- or left-handed. If he starts with the genes or other early physiological influences that predispose him to the left hand, that branch of the developmental channel can be viewed as cutting the more deeply. If no social pressure is exerted the ball will in most cases roll on down into the channel for left-handedness. But if parents train the child to use the right hand, the ball can be nudged into the shallower channel for right-handedness. The landscape for schizophrenia is a broader network of anastomosing channels, more difficult to trace, and the ball's course is only statistically predictable.

The landscape is just a metaphor, and it is certainly inadequate for the most complex phenomena, but it focuses on a crucial truth about human social behavior. If we are to gain full understanding of its determination, each behavior must be treated separately and traced, to some extent, as a developmental process leading from the genes to the final product.

Some forms will prove more susceptible to this mode of analysis than others. The facial expressions displaying the basic emotions of fear, loathing, anger, surprise, and happiness appear to be invariant traits of all human beings. Paul Ekman, a psychologist, took photographs of Americans acting out these emotions. He also photographed stone-age tribesmen as they told stories during which the same feelings were expressed. When members of one of the cultures were then shown the portraits from the other, they interpreted the meanings of the facial expressions with a better than eighty percent accuracy. Irenäus Eibl-Eibesfeldt, traveling to remote communities around the world, has made motion pictures of people as they communicate by gestures and facial expressions. In order to prevent them from being self-conscious, he photographs them through a prism set over the camera lens, an adjustment that permits him to face away from his subject at right angles. Eibl-Eibesfeldt has documented a rich repertory of signals that are widely or even universally distributed through both literate and preliterate cultures. One relatively unfamiliar example is the eyebrow flash—a sudden, mostly unconscious lifting of the eyebrows used as part of a friendly greeting.

Another example of a universal signal being newly studied by human ethologists is the smile, which might qualify as an instinct in a virtually zoological sense. The smile appears on the infant's face between two and four months of age and immediately triggers a more abundant share of parental love and affection. In the terminology of the zoologist, it is a social releaser, an inborn and relatively invariant signal that mediates a basic social relationship. Melvin J. Konner, an anthropologist, has recently completed a study of the smile and other forms of infant behavior in the !Kung San ("Bushmen") of the Kalahari. As he began his daily observations he was "ready for anything," since the !Kung youngsters are raised under very different conditions from those prevailing in Western cultures. They are delivered alone by their mothers, without anesthetic, kept in almost constant physical contact with their mothers or other nurses during the next several months, held in a vertical position during most of their waking hours, nursed several times an hour for the first three or four years, and trained more rigorously than European and American children to sit, stand, and walk. Yet their smile is identical in form, appears at the same age as in American children, and appears to serve exactly the same functions. Still more convincing is the evidence that blind and even deaf-blind children develop the smile in the absence of any known psychological conditioning that favors it.

The simplest and most automatic of such behaviors may well be genetically hard-wired into the cellular units of the human brain and facial nerves, such that the pattern of contraction of the facial muscles develops during early postnatal development by a chain of physiological events requiring a minimum of learning. Closer investigations in the future are likely to disclose the existence of genetic mutations that affect the form and intensity of the neuromuscular actions. If such exceptionally simple phenomena do occur, their discovery will set the stage for our first entrance into the genetics of human communication.

The imagery of the developmental landscape must be altered subtly as increasing amounts of learning and culture come to prevail on the downward slopes. In the case of language, dress, and the other culturally sensitive categories of behavior, the landscape dissolves into a vast delta of low ridges and winding oxbows. Consider in particular the maturation of language. There is evidence that the human mind is innately structured so as to string words together in certain arrangements and not others. According to Noam Chomsky and some other psycholinguists, this "deep grammar" permits a far more rapid acquisition of language than would be possible by simple learning. It is demonstrable by mathematical simulation alone that not enough time exists during childhood to learn English sentences by rote. Young children, unlike the young of any other primates including chimpanzees, possess a fierce drive to acquire speech: they babble, invent words, experiment with meaning, and pick up grammatical rules swiftly and in predictable sequence; they create constructions that anticipate the adult forms and yet differ from them in significant details. Roger Brown, a specialist on child development, has appropriately termed their achievement the "first language." Comparisons between the performances of identical and fraternal twins indicate that variation in the timing of this development depends to some degree on heredity. The upper slope in the developmental field of language is thus a relatively simple and deeply canalized terrain. But the channels of the broad lower slope, where the intricacies of the "second," adult language emerge, make up a shallowly etched network that ramifies in many directions. The outer manifestations of language shift with cultural evolution; they *are* to a large degree cultural evolution. The subtlest pressures from education and fashion alter vocabulary, emphasis, and tempo.

But what in reality corresponds to the metaphorical ridges and channels? In some cases, behaviorally potent hormones, or other biochemical products prescribed by the genes during the construction of nerve cells, etch the channels. Simple compounds can alter the capacity of the nervous system to function in one way as opposed to another. Of equal importance may be the more distantly removed "learning rules," the steps and procedures based on the action of particular sets of nerve cells by which various forms of learning are achieved.

It is commonplace to think of learning as an all-purpose phenomenon that varies little in principle from one kind of organism to the next. Many of the best psychologists, especially B. F. Skinner and other behaviorists, have held stubbornly to the view that most kinds of behavior are shaped by a few elementary forms of learning. By placing animals in simplified laboratory environments, where stimulation can be strictly controlled, the general laws governing learning will be revealed. "The general topography of operant behavior is not important," Skinner wrote in 1938, "because most if not all specific operants are conditioned. I suggest that the dynamic properties of operant behavior may be studied with a single reflex." In his influential book *Beyond Freedom and Dignity*, Skinner argued that once these laws are well understood, they can be used to train human beings to lead happier, more enriched lives. The culture can first be designed by the wisest members of society, and then children fitted painlessly to it.

These are powerful ideas, with seductive precedents in the physical sciences, and they have resulted in substantial advances in the study of animal and human behavior. The central idea of the philosophy of behaviorism, that behavior and the mind have an entirely materialist basis subject to experimental analysis, is fundamentally sound. Nevertheless, the underlying assumptions of simplicity and equipotentiality in learning have crumbled. In their place has emerged a picture of the existence of many peculiar types of learning that conform to no general law except, perhaps, evolution by natural selection. The learning potential of each species appears to be fully programmed by the structure of its brain, the sequence of release of its hormones, and, ultimately, its genes. Each animal species is "prepared" to learn certain stimuli, barred from learning others, and neutral with respect to still others. For example, adult herring gulls quickly learn to distinguish their newly hatched chicks but never their own eggs, which are nevertheless just as visually distinct. The newborn kitten is blind, barely able to crawl on its stomach, and generally helpless. Nevertheless, in the several narrow categories in which it must perform in order to survive, it is endowed with an advanced ability to learn. Using smell alone, it learns in less than one day to crawl short distances to the spot where it can expect to find the nursing mother. With the aid of either odor or touch the kitten memorizes the route along the mother's belly to its own preferred nipple. In laboratory tests it quickly comes to tell one artificial nipple from another by minor differences in texture.

Even more impressive examples have been discovered. Each year indigo buntings migrate between their breeding grounds in eastern North America and their wintering grounds in South America. Like many of our other native birds they travel at night. After leaving the nest, young buntings are prepared to learn the north star and circumpolar constellations, which they proceed to do quickly and automatically. They are inhibited from learning the other constellations. When domestic chicks are given a mild electric shock at the beak while drinking water and are simultaneously given a visual stimulus such as a flash of light, they afterward avoid the visual stimulus, but they do not learn to avoid an auditory stimulus, a clicking sound, in the same way. The reverse is true when the shock is administered to the feet; that is, the chick is prepared to learn sound but not visual cues. This symmetry may seem odd at first but is actually a precise survival rule for a small-brained animal. The chick's procedure can be summarized in the following simple formula: learn the things you can see that affect the head and the things you can hear that affect the feet.

So some of the more rigid forms of animal instinct can be based on idiosyncratic forms of prepared learning. But is human learning prepared? Certainly not in the same robotic fashion as the responses of birds and blind kittens. We like to think that given enough time and will power we can learn anything. Yet constraints exist. We have to concede that there are sharp limits in quantity and complexity to what can be mastered even by geniuses and professional mnemonists, and that everyone acquires certain mental skills far more easily than others. Of still greater significance, children acquire skills and emotions by schedules that are difficult to alter. Switzerland's eminent developmental psychologist, Jean Piaget, has spent a lifetime charting the often surprising stages children pass through in their more purely intellectual growth. The mind follows parallel but tightly coupled tracks in elaborating intentional movements, concepts of meaning and causality, space, time, imitation, and play. Its very conception of reality shifts step by step as the reflex-dominated infant changes into the egocentric and then sociable child. From single-minded efforts to move objects the child's activity grows into a detached reflection on the movements themselves. The objects are first perceived as unique entities and then as members of groups to be classified with the aid of visual symbols and names. Piaget, who was originally trained as a biologist, views intellectual development as an interaction of an inherited genetic program with the environment. It is no coincidence that he calls this conception "genetic epistemology," in effect the study of the hereditary unfolding of understanding.

In his important works *Attachment* and *Separation*, John Bowlby has traced comparable steps in the formation of emotional bonds by which the child creates a complex social world around its parents over a period of months. Lawrence Kohlberg has identified a relatively tight order of Piagetian stages in the growth of moral codes, while psycholinguists have proved that young children acquire language by a time table too precise and too short to be explainable by simple memorization. Considering these accomplishments together, one gains the impression of a social world too complex to be constructed by random learning processes in a lifetime.

So the human mind is not a tabula rasa, a clean slate on which experience draws intricate pictures with lines and dots. It is more accurately described as an autonomous decision-making instrument, an alert scanner of the environment that approaches certain kinds of choices and not others in the first place, then innately leans toward one option as opposed to others and urges the body into action according to a flexible schedule that shifts automatically and gradually from infancy into old age. The accumulation of old choices, the memory of them, the reflection on those to come, the re-experiencing of emotions by which they were engendered, all constitute the mind. Particularities in decision making distinguish one human being from another. But the rules followed are tight enough to produce a broad overlap in the decisions taken by all individuals and hence a convergence powerful enough to be labelled human nature.

It is possible to estimate roughly the relative strictness of the controls on various categories of behavior. Genetic studies based on the comparison of identical and fraternal twins suggest that primary mental abilities and perceptual and motor skills are the most influenced by heredity, while personality traits are the least influenced. If this important result is confirmed by additional studies, the inference to be drawn is that the abilities needed to cope with relatively invariant problems in the physical environment develop along narrow channels, while the qualities of personality, which represent adjustments to the rapidly shifting social environment, are more malleable.

Other correlations of wide significance are suggested by the evolutionary hypothesis. The less rational but more important the decision-making process, for example, the more emotion should be expended in conducting it. The biologist can restate the relationship as follows: much of mental development consists of steps that must be taken quickly and automatically to insure survival and reproduction. Because the brain can be guided by rational calculation only to a limited degree, it must fall back on the nuances of pleasure and pain mediated by the limbic system and other lower centers of the brain.

We can search among the unconscious, emotion-laden learning rules for the kind of behavior most directly influenced by genetic evolution. Consider the phobias. Like many examples of animal learning, they originate most frequently in childhood and are deeply irrational, emotionally colored, and difficult to eradicate. It seems significant that they are most often evoked by snakes, spiders, rats, heights, close spaces, and other elements that were potentially dangerous in our ancient environment, but only rarely by modern artifacts such as knives, guns, and electrical outlets. In early human history phobias might have provided the extra margin needed to insure survival: better to crawl away from a cliff, nauseated by fear, than to walk its edge absent-mindedly.

The incest taboo is an example of another major category of primed learning. As the anthropologists Lionel Tiger and Robin Fox have pointed out, the taboo can be regarded as simply a special case of the more general rule of the precluding of bonds. When two persons form one kind of strong bond between themselves, they find it emotionally difficult to join in certain other kinds. Teachers and students are slow to become colleagues even after the students surpass their mentors; mothers and daughters seldom change the tone of their original relationship. And incest taboos are virtually universal in human cultures because fathers and daughters, mothers and sons, and brothers and sisters find their primary bonds to be nearly all-exclusive. People, in short, are deterred from learning the precluded bonds.

Conversely, people are prepared to learn the genetically most advantageous relationships. The processes of sexual pairbonding vary greatly among cultures, but they are everywhere steeped in emotional feeling. In cultures with a romantic tradition, the attachment can be rapid and profound, creating love beyond sex which, once experienced, permanently alters the adolescent mind. Description of this part of human ethology is the refined specialty of poets, as we see in the remarkable expression by James Joyce:

> A girl stood before him in midstream, alone and still, gazing out to sea. She seemed like one whom magic had changed into the likeness of a strange and beautiful seabird. Her long slender bare legs were delicate as a crane's and pure save where an emerald trail of seaweed had fashioned itself as a sign upon the flesh . . . Her long fair hair was girlish: and girlish, and touched with the wonder of mortal beauty, her face . . . When she felt his presence and the worship of his eyes her eyes turned to him in quiet sufferance of his gaze, without shame or wantonness . . . Her image had passed into his soul for ever and no word had broken the silence of his ecstasy. (*A Portrait of the Artist as a Young Man*)

Prepared learning is logically sought in the other turning points of the life cycle at which our deepest feelings are fixed. Human beings have a strong tendency, for example, to manufacture thresholds across which they step ritualistically from one existence to another. Culture elaborates the rites of passage—initiation, marriage, confirmation, and inauguration—in ways perhaps affected by still hidden biological prime movers. In all periods of life there is an equally powerful urge to dichotomize, to classify other human beings into two artificially sharpened categories. We seem able to be fully comfortable only when the remainder of humanity can be labelled as members versus nonmembers, kin versus nonkin, friend versus foe. Erik Erikson has written on the proneness of people everywhere to perform pseudospeciation, the reduction of alien societies to the status of inferior species, not fully human, who can be degraded without conscience. Even the gentle San of the Kalahari call themselves the !Kung—*the* human beings. These and other of the all-too-human predispositions make complete sense only when valuated in the coinage of genetic advantage. Like the appealing springtime songs of male birds that serve to defend territories and to advertise aggression, they possess an esthetic whose true, deadly meaning is at first concealed from our conscious minds.

Chapter 4. Emergence

If biology is destiny, as Freud once told us, what becomes of free will? It is tempting to think that deep within the brain lives a soul, a free agent that takes account of the body's experience but travels around the cranium on its own accord, reflecting, planning, and pulling the levers of the neuromotor machinery. The great paradox of determinism and free will, which has held the attention of the wisest of philosophers and psychologists for generations, can be phrased in more biological terms as follows: if our genes are inherited and our environment is a train of physical events set in motion before we were born, how can there be a truly independent agent within the brain? The agent itself is created by the interaction of the genes and the environment. It would appear that our freedom is only a self-delusion.

In fact, this may be so. It is a defensible philosophical position that at least some events above the atomic level are predictable. To the extent that the future of objects can be foretold by an intelligence which itself has a material basis, they are determined—but only within the conceptual world of the observing intelligence. And insofar as they can make decisions of their own accord—whether or not they are determined—they possess free will. Consider the flip of a coin and the extent of the coin's freedom. On first thought nothing could seem less subject to determinism; coin flipping is the classic textbook example of a random process. But suppose that for some reason we decided to bring all the resources of modern science to bear on a single toss. The coin's physical properties are measured to the nearest picogram and micron, the muscle physiology and exact contours of the flipper's thumb are analyzed, the air currents of the room charted, the microtopography and resiliency of the floor surface mapped. At the moment of release, all of this information, plus the instantaneously recorded force and angle of the flip, are fed into a computer. Before the coin has spun through more than a few revolutions, the computer reports the expected full trajectory of the coin and its final resting position at heads or tails. The method is not perfect, and tiny errors in the initial conditions of the flip can be blown up during computation into an error concerning the outcome. Nevertheless, a series of computer-aided predictions will probably be more accurate than a series of guesses. To a limited extent, we can know the destiny of the coin.

An interesting exercise, one can reply, but not entirely relevant, because the coin has no mind. This deficiency can be remedied stepwise, by first selecting a circumstance of intermediate complexity. Let the object propelled into the air be an insect, say a honeybee. The bee has a memory. It can think in a very limited way. During its very short life—it will die of old age at fifty days—it has learned the time of day, the location of its hive, the odor of its nestmates, and the location and quality of up to five flower fields. It will respond vigorously and erratically to the flick of the scientist's hand that knocks it loose. The bee appears to be a free agent to the uninformed human observer, but again if we were to concentrate all we know about the physical properties of thimble-sized objects, the nervous system of insects, the behavioral peculiarities of honeybees, and the personal history of this particular bee, and if the most advanced computational techniques were again brought to bear, we might predict the flight path of the bee with an accuracy that exceeds pure chance. To the circle of human observers watching the computer read-out, the future of the bee is determined to some extent. But in her own "mind" the bee, who is isolated permanently from such human knowledge, will always have free will.

When human beings ponder their own central nervous systems, they appear at first to be in the same position as the honeybee. Even though human behavior is enormously more complicated and variable than that of insects, theoretically it can be specified. Genetic constraints and the restricted number of environments in which human beings can live limit the array of possible outcomes substantially. But only techniques beyond our present imagining could hope to achieve even the short-term prediction of the detailed behavior of an individual human being, and such an accomplishment might be beyond the capacity of any conceivable intelligence. There are hundreds or thousands of variables to consider, and minute degrees of imprecision in any one of them might easily be magnified to alter the action of part or all of the mind. Furthermore, an analog of the Heisenberg uncertainty principle in subatomic physics is at work here on a grander scale: the more deeply the observer probes the behavior, the more the behavior is altered by the act of probing and the more its very meaning depends on the kinds of measurements chosen. The will and destiny of the watcher is linked to that of the person watched. Only the most sophisticated imaginable monitoring devices, capable of recording vast numbers of internal nervous processes simultaneously and from a distance, could reduce the interaction to an acceptably low level. Thus because of mathematical indeterminancy and the uncertainty principle, it may be a law of nature that no nervous system is capable of acquiring enough knowledge to significantly predict the future of any other intelligent system in detail. Nor can intelligent minds gain enough self-knowledge to know their own future, capture fate, and in this sense eliminate free will.

An equally basic difficulty in making a forecast of an activity as complicated as the human mind lies in the transformations through which raw data reach the depths of the brain. Vision, for example, begins its journey when the radiant energy of light triggers electrical activity in the approximately one hundred million primary light receptor cells that comprise the retina. Each cell records the level of brightness (or color) that touches it in each instant of time; the image transmitted through the lens is thus picked up as a pattern of electrical signals in the manner of a television camera. Behind the retina a million or so ganglion cells receive the signals and process them by a form of abstraction. Each cell receives information from a circular cluster of primary receptors in the retina. When a light-dark contrast of sufficient intensity divides the retinal cluster, the ganglion cell is activated. This information is then passed on to a region of the cerebral cortex low in the back of the head, where special cortical nerve cells reinterpret it. Each cortical cell is activated by a group of subordinate ganglion cells. It responds with electrical activity if the pattern in which the ganglion cells are discharged reflects a straight line edge of one or the other of three particular orientations: horizontal, vertical, or oblique. Other cortical cells, carrying the abstraction still further, respond either to the ends of straight lines or to corners.

The mind might well receive all of its information, originating from both outside and inside the body, through such coding and abstracting processes. Consciousness consists of immense numbers of simultaneous and coordinated, symbolic representations by the participating neurons of the brain's neocortex. Yet to classify consciousness as the action of organic machinery is in no way to underestimate its power. In Sir Charles Sherrington's splendid metaphor, the brain is an "enchanted loom where millions of flashing shuttles weave a dissolving pattern." Since the mind recreates reality from the abstractions of sense impressions, it can equally well simulate reality by recall and fantasy. The brain invents stories and runs imagined and remembered events back and forth through time: destroying enemies, embracing lovers, carving tools from blocks of steel, travelling easily into the realms of myth and perfection.

The self is the leading actor in this neural drama. The emotional centers of the lower brain are programmed to pull the puppeteer's strings more carefully whenever the self steps onto the stage. But granted that our deepest feelings are about ourselves, can this preoccupation account for the innermost self — the *soul* — in mechanistic terms? The cardinal mystery of neurobiology is not self-love or dreams of immortality but intentionality. What is the prime mover, the weaver who guides the flashing shuttles? Too simple a neurological approach can lead to an image of the brain as a Russian doll: in the same way that we open one figure after another to reveal a smaller figure until nothing remains, our research resolves one system of neuron circuits after another into smaller subcircuits until only isolated cells remain. At the opposite extreme too complex a neurological model can lead back to a vitalistic metaphysics, in which properties are postulated that cannot be translated into neurons, circuits, or any other physical units.

The compromise solution might lie in recognizing what cognitive psychologists call schemata or plans. A schema is a configuration within the brain, either inborn or learned, against which the input of the nerve cells is compared. The matching of the real and expected patterns can have one or the other of several effects. The schema can contribute to a person's mental "set," the screening out of certain details in favor of others, so that the conscious mind perceives a certain part of the environment more vividly than others and is likely to favor one kind of decision over another. It can fill in details that are missing from the actual sensory input and create a pattern in the mind that is not entirely present in reality. In this way the gestalt of objects — the impression they give of being a square, a face, a tree, or whatever — is aided by the taxonomic powers of the schemata. The frames of reference serve to coordinate movement of the entire body by creating an awareness and automatic control of its moveable parts. The coupling of sensory input and these frames is dramatically illustrated when a limb has been immobilized by injury and is put back into use. A psychologist, Oliver Sacks, has described his own sensations when trying to take a first step after long recuperation from a leg injury:

> I was suddenly precipitated into a sort of perceptual delirium, an incontinent bursting-forth of representations and images unlike anything I had ever experienced before. Suddenly my leg and the ground before me seemed immensely far away, then under my nose, then bizarrely tilted or twisted one way or another. These wild perceptions (or perceptual hypotheses) succeeded one another at the rate of several per second, and were generated in an involuntary and incalculable way. By degrees they came less erratic and wild, until finally, after perhaps five minutes and a thousand such flashes, a plausible image of the leg was achieved. With this the leg suddenly felt *mine* and real again, and I was forthwith able to walk.

Most significantly of all, schemata within the brain could serve as the physical basis of will. An organism can be guided in its actions by a feedback loop: a sequence of messages from the sense organs to the brain schemata back to the sense organs and on around again until the schemata "satisfy" themselves that the correct action has been completed. The mind could be a republic of such schemata, programmed to compete among themselves for control of the decision centers, individually waxing or waning in power in response to the relative urgency of the physiological needs of the body being signaled to the conscious mind through the brain stem and midbrain. Will might be the outcome of the competition, requiring the action of neither a "little man" nor any other external agent. There is no proof that the mind works in just this way. For the moment suffice it to note that the basic mechanisms do exist; feedback loops, for example, control most of our automatic behavior. It is entirely possible that the will — the soul, if you wish — emerged through the evolution of physiological mechanisms. But, clearly, such mechanisms are far more complex than anything else on earth.

So, for the moment, the paradox of determinism and free will appears not only resolvable in theory, it might even be reduced in status to an empirical problem in physics and biology. We note that even if the basis of mind is truly mechanistic, it is very unlikely that any intelligence could exist with the power to predict the precise actions of an individual human being, as we might to a limited degree chart the path of a coin or the flight of a honeybee. The mind is too complicated a structure, and human social relations affect its decisions in too intricate and variable a manner, for the detailed histories of individual human beings to be predicted in advance by the individuals affected or by other human beings. You and I are consequently free and responsible persons in this fundamental sense.

And yet our behavior is partially determined in a second and weaker sense. If the categories of behavior are made broad enough, events can be predicted with confidence. The coin will spin and not settle on its edge, the bee will fly around the room in an upright position, and the human being will speak and conduct a wide range of social activities characteristic of the human species. Moreover, the statistical properties of *populations* of individuals can be specified. In the case of spinning coins, there is no need for computers and other paraphernalia to make statistical projections exact; the binomial distribution and arc-sine laws governing their behavior can be easily written on the back of an envelope, and these mathematical formulas are rich with useful information. At another level, entomologists have produced detailed characterizations of the averaged flight patterns of honeybees to flowers. They know in advance the statistical properties of the waggle dance the bees will perform to convey the location of the flowers to nestmates. They have measured the timing and precise distribution of errors made by bees acting on that information.

To a lesser and still unknown degree the statistical behavior of human societies might be predicted, given a sufficient knowledge of human nature, the histories of the societies, and their physical environment.

Genetic determination narrows the avenue along which further cultural evolution will occur. There is no way at present to guess how far that evolution will proceed. But its past course can be more deeply interpreted and perhaps, with luck and skill, its approximate future direction can be charted. The psychology of individuals will form a key part of this analysis. Despite the imposing holistic traditions of Durkheim in sociology and Radcliffe-Brown in anthropology, cultures are not superorganisms that evolve by their own dynamics. Rather, cultural change is the statistical product of the separate behavioral responses of large numbers of human beings who cope as best they can with social existence.

When societies are viewed strictly as populations, the relationship between culture and heredity can be defined more precisely. Human social evolution proceeds along a dual track of inheritance: cultural and biological. Cultural evolution is Lamarckian and very fast, whereas biological evolution is Darwinian and usually very slow.

Lamarckian evolution would proceed by the inheritance of acquired characteristics, the transmission to offspring of traits acquired during the lifetime of the parent. When the French biologist Jean

Baptiste de Lamarck proposed the idea in 1809, he believed that biological evolution occurred in just such a manner. He suggested, for example, that when giraffes stretch their necks to feed on taller trees, their offspring acquire longer necks even without such an effort; and when storks stretch their legs to keep their bellies dry, their offspring inherit longer legs in the same direct way. Lamarckism has been entirely discounted as the basis of biological evolution, but of course it is precisely what happens in the case of cultural evolution.

The great competing theory of evolution, that entire populations are modified by natural selection, was first put in convincing form by Charles Darwin, in 1859. Individuals within populations vary in their genetic composition and thus in their ability to survive and reproduce. Those that are most successful pass more hereditary material to the next generation, and as a result the population as a whole progressively changes to resemble the successful types. Individual giraffes, by the theory of natural selection, differ from one another in the hereditary capacity to grow long necks. Those that do develop the longest necks feed more and leave the higher proportion of offspring; as a consequence the average neck length of the giraffe population increases over many generations. If, in addition, genetic mutations occurring from time to time affect neck length, the process of evolution can continue indefinitely.

Darwinism has been established as the prevailing mode of biological evolution in all kinds of organisms, including man. Because it is also far slower than Lamarckian evolution, biological evolution is always quickly outrun by cultural change. Yet the divergence cannot become too great, because ultimately the social environment created by cultural evolution will be tracked by biological natural selection. Individuals whose behavior has become suicidal or destructive to their families will leave fewer genes than those genetically less prone to such behavior. Societies that decline because of a genetic propensity of its members to generate competitively weaker cultures will be replaced by those more appropriately endowed. I do not for a moment ascribe the relative performances of modern societies to genetic differences, but the point must be made: there is a limit, perhaps closer to the practices of contemporary societies than we have had the wit to grasp, beyond which biological evolution will begin to pull cultural evolution back to itself.

And more: individual human beings can be expected to resist too great a divergence between the two evolutionary tracks. Somewhere in the mind, as Lionel Trilling said in *Beyond Culture*, "there is a hard, irreducible, stubborn core of biological urgency, and biological necessity, and biological *reason*, that culture cannot reach and that reserves the right, which sooner or later it will exercise, to judge the culture and resist and revise it."

Such biological refractoriness is illustrated by the failure of slavery as a human institution. Orlando Patterson, a sociologist at Harvard University, has made a systematic study of the history of slave societies around the world. He has found that true, formalized slavery passes repeatedly through approximately the same life cycle, at the end of which the peculiar circumstances stemming from its origin together with the stubborn qualities of human nature lead to its destruction.

Large-scale slavery begins when the traditional mode of production is dislocated, usually due to warfare, imperial expansion, and changes in basic crops, which in turn induces the rural free poor to migrate into the cities and newly opened colonial settlements. At the imperial center, land and capital fall increasingly under the monopoly of the rich, while citizen labor grows scarcer. The territorial expansion of the state, by making the enslavement of other peoples profitable, temporarily solves the economic problem. Were human beings then molded by the new culture, were they to behave like the red *Polyergus* ants for which slavery is an automatic response, slave societies might become permanent. But the qualities that we recognize as most distinctively mammalian — and human — make such a transition impossible. The citizen working class becomes further divorced from the means of production because of their aversion to the low status associated with common labor. The slaves, meanwhile, attempt to maintain family and ethnic relationships and to piece together the shards of their old culture. Where the effort succeeds, many of them rise in status and alter their position from its original, purely servile form. Where self-assertion fails because it is suppressed, reproduction declines and large numbers of new slaves must be imported in each generation. The rapid turnover has a disintegrating effect on the culture of slaves and masters alike. Absenteeism rises as the slave owners attempt to spend more of their time in the centers of their own culture. Overseers come increasingly into control. Inefficiency, brutality, revolt, and sabotage increase, and the system spirals slowly downward.

Slave-supported societies, from ancient Greece and Rome to medieval Iraq and eighteenth century Jamaica, have had many other flaws, some of which might have been fatal. But the institution of slavery alone has been enough to ordain the spectacular sweep of their life cycle. "Their ascent to maturity is rapid," Patterson writes, "their period of glory short, and their descent to oblivion ostentatious and mightily drawn out."

The fact that slaves under great stress insist on behaving like human beings instead of slave ants, gibbons, mandrills, or any other species, is one of the reasons I believe that the trajectory of history can be plotted ahead, at least roughly. Biological constraints exist that define zones of improbable or forbidden entry. In suggesting the possibility of a certain amount of revealed destiny (a theme that will be elaborated in the final chapter), I am well aware that it is within human capacity to legislate any hypothetical course of history as opposed to another. But even if the power of self-determination is turned full on, the energy and materials crises solved, old ideologies defeated, and hence all societal options laid open, there are still only a few directions we will want to take. Others may be tried, but they will lead to social and economic perturbations, a decline in the quality of life, resistance, and retreat.

If it is true that history is guided to a more than negligible extent by the biological evolution that preceded it, valuable clues to its course can be found by studying the contemporary societies whose culture and economic practices most closely approximate those that prevailed during prehistory. These are the hunter-gatherers: the Australian aboriginals, Kalahari San, African pygmies, Andaman Negritos, Eskimos, and other peoples who depend entirely on the capture of animals and harvesting of free-growing plant material. Over one hundred such cultures still survive. Few contain over ten thousand members, and almost all are in danger of assimilation into surrounding cultures or outright extinction. Anthropologists, being fully aware of the great theoretical significance of these primitive cultures, are now pitted in a race against time to record them before they disappear.

Hunter-gatherers share many traits that are directly adaptive to their rugged way of life. They form bands of a hundred or less that roam over large home ranges and often divide or rejoin each other in the search for food. A group comprising twenty-five individuals typically occupies between one thousand and three thousand square kilometers, an area comparable to the home range of a wolf pack of the same size but a hundred times greater than what a troop of exclusively vegetarian gorillas would occupy. Parts of the ranges are sometimes defended as territories, especially those containing rich and reliable sources of food. Intertribal aggression, escalating in some cultures to limited warfare, is common enough to be regarded as a general characteristic of hunter-gatherer social behavior.

The band is, in reality, an extended family. Marriage is arranged within and between bands by negotiation and ritual, and the complex kinship networks that result are objects of special classifications and strictly enforced rules. The men of the band, while leaning toward mildly polygamous arrangements, make substantial investments of time in rearing their offspring. They are also protective of their investments. Murder, which is as common per capita as in most American cities, is most often committed in response to adultery and during other disputes over women.

The young pass through a long period of cultural indoctrination during which the focus of their activities shifts gradually from the mother to age and peer groups. Their games promote physical skill but not strategy, and simulate in relatively unorganized and rudimentary form the adult roles the children will later adopt.

A strong sexual division of labor prevails in every facet of life. Men are dominant over women only in the sense of controlling certain tribal functions. They preside at councils, decide the forms of rituals, and control exchanges with neighboring groups. Otherwise, the ambience is informal and egalitarian by comparison with the majority of economically more complex societies. Men hunt and women gather. Some overlap of these roles is common, but the overlap becomes less when game is large and pursued over long distances. Hunting usually has an important but not overwhelming role in the economy. In his survey of sixty-eight hunter-gatherer societies, the anthropologist Richard B. Lee has found that on average only about one-third of the diet consists of fresh meat. Even so, this food contains the richest, most desired source of proteins and fats, and it usually confers the most prestige to its owners.

Among the many carnivores patrolling the natural environment, primitive men are unusual in capturing prey larger than themselves. Although many of the animals they pursue are small — lying within the combined size range of mice, birds, and lizards — no great creature is immune. Walruses, giraffes, kudu, and elephants fall to the snares and hand-carved weapons of the hunters. The only other mammalian carnivores that take outsized prey are lions, hyenas, wolves, and African wild dogs. Each of these species has an exceptionally advanced social life, prominently featuring the pursuit of prey in coordinated packs. The two traits, large prey size and social hunting, are unquestionably linked. Lions, which are the only social members of the cat family, double their catch when hunting in prides. In addition they are able to subdue the largest and most difficult prey, including giraffes and adult male buffalos, which are almost invulnerable to single predators. Primitive men are ecological analogs of lions, wolves, and hyenas. Alone among the primates, with the marginal exception of the chimpanzees, they have adopted pack hunting in the pursuit of big game. And they resemble four-footed carnivores more than other primates by virtue of habitually slaughtering surplus prey, storing food, feeding solid food to their young, dividing labor, practicing cannibalism, and interacting aggressively with competing species. Bones and stone tools dug from ancient campsites in Africa, Europe, and Asia indicate that this way of life persisted for a million years or longer and was abandoned in most societies only during the last few thousands of years. Thus the selection pressures of hunter-gatherer existence have persisted for over 99 percent of human genetic evolution.

This apparent correlation between ecology and behavior brings us to the prevailing theory of the origin of human social behavior. It consists of a series of interlocking reconstructions that have been fashioned from bits of fossil evidence, extrapolations back through time from hunter-gatherer societies, and comparisons with other living primate species. The core of the theory is what I referred to in my earlier book *Sociobiology* as the *autocatalysis model*. Autocatalysis is a term that originated in chemistry; it means any process that increases in speed according to the amount of the products it has created. The longer the process runs, the greater its speed. By this conception the earliest men or man-apes started to walk erect when they came to spend most or all of their time on the ground. Their hands were freed, the manufacture and handling of artifacts were made easier, and intelligence grew as the tool-using habit improved. With mental capacity and the tendency to use artifacts increasing through mutual reinforcement, the entire materials-based culture expanded. Now the species moved onto the dual track of evolution: genetic evolution by natural selection enlarged the capacity for culture, and culture enhanced the genetic fitness of those who made maximum use of it. Cooperation during hunting was perfected and provided a new impetus for the evolution of intelligence, which in turn permitted still more sophistication in tool using, and so on through repeated cycles of causation. The sharing of game and other food contributed to the honing of social skills. In modern hunter-gatherer bands, it is an occasion for constant palavering and maneuvering. As Lee said of the !Kung San,

> The buzz of conversation is a constant background to the camp's activities: there is an endless flow of talk about gathering, hunting, the weather, food distribution, gift giving, and scandal. No !Kung is ever at a loss for words, and often two or three people will hold forth at once in a single conversation, giving the listeners a choice of channels to tune in on. A good proportion of this talk in even the happiest of camps verges on argument. People argue about improper food division, about breaches of etiquette, and about failure to reciprocate hospitality and gift giving . . . Almost all the arguments are *ad hominem*. The most frequent accusations heard are of pride, arrogance, laziness, and selfishness.

The natural selection generated by such exchanges might have been enhanced by the more sophisticated social behavior required by the female's nearly continuous sexual accessibility. Because a high level of cooperation exists within the band, sexual selection would be linked with hunting prowess, leadership, skill at tool making, and other visible attributes that contribute to the strength of the family and the male band. At the same time aggressiveness would have to be restrained and the phylogenetically ancient forms of overt primate dominance replaced by complex social skills. Young males would find it profitable to fit into the group by controlling their sexuality and aggression and awaiting their turn at leadership. The dominant male in these early hominid societies was consequently most likely to possess a mosaic of qualities that reflect the necessities of compromise. Robin Fox has suggested the following portrait: "Controlled, cunning, cooperative, attractive to the ladies, good with the children, relaxed, tough, eloquent, skillful, knowledgeable and proficient in self-defense and hunting." Because there would have been a continuously reciprocating relationship between the more sophisticated social traits and breeding success, social evolution could continue indefinitely without additional selective pressures from the environment.

At some point, possibly during the transition from the more primitive *Australopithecus* man-apes to the earliest true men, the autocatalysis carried the evolving populations to a new threshold of competence, at which time the hominids were able to exploit the sivatheres, elephants, and other large herbivorous animals teeming around them on the African plains. Quite possibly the process began when the hominids learned to drive big cats, hyenas, and other carnivores away from their kills. In time the hominids became the primary hunters and were forced to protect their prey from other predators and scavengers.

Child care would have been improved by close social bonding between individual males, who left the domicile to hunt larger game, and individual females, who kept the children and conducted most of the foraging for vegetable food. In a sense, love was added to sex. Many of the peculiar details of human sexual behavior and domestic life flow easily from this basic division of labor. But such details are not essential to the autocatalysis model. They are appended to the evolutionary story only because they are displayed by virtually all hunter-gatherer societies.

Autocatalytic reactions never expand to infinity, and biological processes themselves normally change through time to slow growth and eventually bring it to a halt. But almost miraculously, this has not yet happened in human evolution. The increase in brain size and refinement of stone artifacts point to an unbroken advance in mental ability over the last two to three million years. During this crucial period the brain evolved in either one great surge or a series of alternating surges and plateaus. No organ in the history of life has grown faster. When true men diverged from the ancestral man-apes, the brain added one cubic inch — about a tablespoonful — every hundred thousand years. The rate was maintained until about one quarter of a million years ago, when, at about the time of the appearance of the modern species *Homo sapiens*, it tapered off. Physical growth was then supplanted by an increasingly prominent cultural evolution. With the appearance of the Mousterian tool culture of the Neanderthal man some seventy-five thousand years ago, cultural change gathered momentum, giving rise in Europe to the Upper Paleolithic culture of Cro-Magnon man about forty thousand years before the present. Starting about ten thousand years ago agriculture was invented and spread, populations increased enormously in density, and the primitive hunter-gatherer bands gave way locally to the relentless growth of tribes, chiefdoms, and states. Finally, after A.D. 1400 European-based civilization shifted gears again, and the growth of knowledge and technology accelerated to world-altering levels.

There is no reason to believe that during this final sprint to the space age there has been a cessation in the evolution of either mental capacity or the predilection toward special social behaviors. The theory of population genetics and experiments on other organisms show that substantial changes can occur in the span of less than 100 generations, which for man reaches back only to the time of the Roman Empire. Two thousand generations, roughly the time since typical *Homo sapiens* invaded Europe, is enough time to create new species and to mold their anatomy and behavior in major ways. Although we do not know how much mental evolution has actually occurred, it would be premature to assume that modern civilizations have been built entirely on genetic capital accumulated during the long haul of the Ice Age.

That capital is nevertheless very large. It seems safe to assume that the greater part of the changes that transpired in the interval from the hunter-gatherer life of forty thousand years ago to the first glimmerings of civilization in the Sumerian city states, and virtually all of the changes from Sumer to Europe, were created by cultural rather than genetic evolution. The question of interest, then, is the extent to which the hereditary qualities of hunter-gatherer existence have influenced the course of subsequent cultural evolution.

I believe that the influence has been substantial. In evidence is the fact that the emergence of civilization has everywhere followed a definable sequence. As societies grew in size from the tiny hunter-gatherer bands, the complexity of their organization increased by the addition of features that appeared in a fairly consistent order. As band changed to tribe, true male leaders appeared and gained dominance, alliances between neighboring groups were strengthened and formalized, and rituals marking the changes of season became general. With still denser populations came the attributes of generic chiefdom: the formal distinction of rank according to membership in families, the hereditary consolidation of leadership, a sharper division of labor, and the redistribution of wealth under the control of the ruling elite. As chiefdoms gave rise in turn to cities and states, these basic qualities were intensified. The hereditary status of the elite was sanctified by religious beliefs. Craft specialization formed the basis for stratifying the remainder of society into classes. Religion and law were codified, armies assembled, and bureaucracies expanded. Irrigation systems and agriculture were perfected, and as a consequence populations grew still denser. At the apogee of the state's evolution, architecture was monumental, and the ruling classes were exalted as a pseudospecies. The sacred rites of statehood became the central focus of religion.

The similarities between the early civilizations of Egypt, Mesopotamia, India, China, Mexico, and Central and South America in these major features are remarkably close. They cannot be explained away as the products of chance or cultural cross-fertilization. It is true that the archives of ethnography and history are filled with striking and unquestionably important variations in the details of culture, but it is the parallelism in the major features of organization that demands our closest attention in the consideration of the theory of the dual track of human social evolution.

In my opinion the key to the emergence of civilization is *hypertrophy*, the extreme growth of pre-existing structures. Like the teeth of the baby elephant that lengthen into tusks, and the cranial bones of the male elk that sprout into astonishing great antlers, the basic social responses of the hunter-gatherers have metamorphosed from relatively modest environmental adaptations into unexpectedly elaborate, even monstrous forms in more advanced societies. Yet the directions this change can take and its final products are constrained by the genetically influenced behavioral predispositions that constituted the earlier, simpler adaptations of preliterate human beings.

Hypertrophy can sometimes be witnessed at the beginning. One example in its early stages is the subordination of women in elementary cultures. The !Kung San of the Kalahari Desert do not impose sex roles on their children. Adults treat little girls in apparently the

Type of society	Some institutions, in order of appearance	Ethnographic examples	Archaeological examples
STATE	Taxation; Military draft; Bureaucracy; Codified law; Kingship; Stratification; Full-time craft specialization; Elite endogamy; Hereditary leadership; Redistributive economy; Ranked descent groups; Calendric ritual; Pantribal sodalities; Unranked descent groups; Reciprocal economy; Ad hoc ritual; Ephemeral leadership; Egalitarian status; Local group autonomy	France, England, India, U.S.A.	Classic Mesoamerica, Sumer, Shang China, Imperial Rome
CHIEFDOM	(Full-time craft specialization; Elite endogamy; Hereditary leadership; Redistributive economy; Ranked descent groups; Calendric ritual; ...)	Tonga, Hawaii, Kwakiutl, Nootka, Natchez	Gulf Coast Olmec of Mexico (1000 B.C.), Samarran of Near East (5300 B.C.), Mississippian of North America (A.D. 1200)
TRIBE	(Calendric ritual; Pantribal sodalities; Unranked descent groups; Reciprocal economy; Ad hoc ritual; ...)	New Guinea Highlanders, Southwest Pueblos, Sioux	Early Formative of Inland Mexico (1500-1000 B.C.), Prepottery Neolithic of Near East (8000-6000 B.C.)
BAND	(Reciprocal economy; Ad hoc ritual; Ephemeral leadership; Egalitarian status; Local group autonomy)	Kalahari San, Australian Aborigines, Eskimo, Shoshone	Paleoindian and Early Archaic of U.S. and Mexico (10,000-6000 B.C.), Late Paleolithic of Near East (10,000 B.C.)

As societies grew larger, they acquired new institutions in a roughly consistent order. This diagram shows examples from the historical sequence (column on the far right) and existing cultures (second column from the right). (Based on K. V. Flannery.)

same manner as little boys, which is to say with considerable indulgence and permissiveness. Yet, as the anthropologist Patricia Draper found during a special study of child development, small average differences still appear. From the beginning the girls stay closer to home and join groups of working adults less frequently. During play, boys are more likely to imitate the men, and girls are more likely to imitate the women. As the children grow up, these differences lead through imperceptible steps to a still stronger difference in adult sex roles. Women gather mongongo nuts and other plant food and fetch water, usually within a mile of camp, while men range farther in search of game. But !Kung social life is relaxed and egalitarian, and tasks are often shared. Men sometimes gather mongongo nuts or build huts (women's work), with or without their families, and women occasionally catch small game. Both sexual roles are varied and esteemed by all. According to Draper, !Kung women maintain personal control over the food they gather, and in demeanor they are generally "vivacious and self-confident."

In a few localities bands have settled into villages to take up farming. The work is heavier, and for the first time in known !Kung history it has come to be shared to a significant extent by the younger children. The sexual roles are noticeably hardened from early childhood onward. Girls stay even closer to the home than previously in order to care for smaller children and perform household chores. Boys tend herds of domestic animals and protect the gardens from monkeys and goats. By maturity the sexes have diverged far from one another in both way of life and status. The women are more fully domestic, working almost continuously at a multiplicity of tasks in which they are supervised. The men continue to wander freely, taking responsibility for their own time and activities.

So only a single lifetime is needed to generate the familiar pattern of sexual domination in a culture. When societies grow still larger and more complex, women tend to be reduced in influence outside the home, and to be more constrained by custom, ritual, and formal law. As hypertrophy proceeds further, they can be turned literally into chattel, to be sold and traded, fought over, and ruled under a double morality. History has seen a few striking local reversals, but the great majority of societies have evolved toward sexual domination as though sliding along a ratchet.

Most and perhaps all of the other prevailing characteristics of modern societies can be identified as hypertrophic modifications of the biologically meaningful institutions of hunter-gatherer bands and early tribal states. Nationalism and racism, to take two examples, are the culturally nurtured outgrowths of simple tribalism. Where the Nyae Nyae !Kung speak of themselves as perfect and clean and other !Kung people as alien murderers who use deadly poisons, civilizations have raised self-love to the rank of high culture, exalted themselves by divine sanction and diminished others with elaborately falsified written histories.

Even the beneficiaries of the hypertrophy have found it difficult to cope with extreme cultural change, because they are sociobiologically equipped only for an earlier, simpler existence. Where the hunter-gatherer fills at most one or two informal roles out of only several available, his literate counterpart in an industrial society must choose ten or more out of thousands, and replace one set with another at different periods of his life or even at different times of the day. Furthermore, each occupation — the physician, the judge, the teacher, the waitress — is played just so, regardless of the true workings of the mind behind the persona. Significant deviations in performance are interpreted by others as a sign of mental incapacity and unreliability. Daily life is a compromised blend of posturing for the sake of role-playing and of varying degrees of self-revelation. Under these stressful conditions even the "true" self cannot be precisely defined, as Erving Goffman observes.

There is a relation between persons and role. But the relationship answers to the interactive system — to the frame — in which the role is performed and the self of the performer is glimpsed. Self, then, is not an entity half-concealed behind events, but a changeable formula for managing oneself during them. Just as the current situation prescribes the official guise behind which we will conceal ourselves, so it provides where and how we will show through, the culture itself prescribing what sort of entity we must believe ourselves to be in order to have something to show through in this manner.

Little wonder that the identity crisis is a major source of modern neuroticism, and that the urban middle class aches for a return to a simpler existence.

As these various cultural superstructures have proliferated, their true meaning more often than not has become lost to the practitioners. In *Cannibals and Kings*, Marvin Harris has suggested a series of bizarre examples of the way that chronic meat shortages affect the shaping of religious beliefs. While the ancient hunter-gatherers were beset with daily perils and constricting fluctuations in the environments that kept their populations low in density, they could at least count on a relatively high fraction of fresh meat in their diet. Early human beings, as I have said, filled a special ecological niche: they were the carnivorous primates of the African plains. They retained this position throughout the Ice Age as they spread into Europe, Asia, and finally into Australia and the New World. When agriculture permitted the increase of population density, game was no longer abundant enough to provide a sufficient supply of fresh meat, and the rising civilizations either switched to domestic animals or went on reduced rations. But in either case carnivorism remained a basic dietary impulse, with cultural aftereffects that varied according to the special conditions of the environment in which the society evolved.

Ancient Mexico, like most of the forest-invested New World tropics, was deficient in the kind of large game that flourished on the plains of Africa and Asia. Furthermore, the Aztecs and other peoples who built civilizations there failed to domesticate animals as significant sources of meat. As human populations grew thicker in the Valley of Mexico, the Aztec ruling class was still able to enjoy such delicacies as dogs, turkeys, ducks, deer, rabbits, and fish. But animal flesh was virtually eliminated from the diets of the commoners, who were occasionally reduced to eating clumps of spirulina algae skimmed from the surface of Lake Texcoco. The situation was partially relieved by cannibalizing the victims of human sacrifice. As many as fifteen thousand persons a year were being consumed in the Valley of Mexico when Cortez entered. The conquistadors found a hundred thousand skulls stacked in neat rows in the plaza at Xocotlan and another 136 thousand at Tenochtitlán. The priesthood said that human sacrifice was approved by the high gods, and they sanctified it with elaborate rituals performed amid statuary of the gods placed on imposing white temples erected for this purpose. But these trappings should not distract us from the fact that immediately after their hearts had been cut out, the victims were systematically butchered like animals and their parts distributed and eaten. Those favored in the feasts included the nobility, their retainers, and the soldiery, in other words the groups with the greatest political power.

India began from a stronger nutrient base than Mexico and followed a different but equally profound cultural transformation as meat grew scarce. The earlier Aryan invaders of the Gangetic Plain presided over feasts of cattle, horses, goats, buffalo, and sheep. By later Vedic and early Hindu times, during the first millenium B.C., the feasts came to be managed by the priestly caste of Brahmans, who erected rituals of sacrifice around the killing of animals and distributed the meat in the name of the Aryan chiefs and war lords. After 600 B.C., when populations grew denser and domestic animals became proportionately scarcer, the eating of meat was progressively restricted until it became a monopoly of the Brahmans and their sponsors. Ordinary people struggled to conserve enough livestock to meet their own desperate requirements for milk, dung used as fuel, and transport. During this period of crisis, reformist religions arose, most prominently Buddhism and Jainism, that attempted to abolish castes and hereditary priesthoods and to outlaw the killing of animals. The masses embraced the new sects, and in the end their powerful support reclassified the cow into a sacred animal.

So it appears that some of the most baffling of religious practices in history might have an ancestry passing in a straight line back to the ancient carnivorous habits of humankind. Cultural anthropologists like to stress that the evolution of religion proceeds down multiple, branching pathways. But these pathways are not infinite in number; they may not even be very numerous. It is even possible that with a more secure knowledge of human nature and ecology, the pathways can be enumerated and the directions of religious evolution in individual cultures explained with a high level of confidence.

I interpret contemporary human social behavior to comprise hypertrophic outgrowths of the simpler features of human nature joined together into an irregular mosaic. Some of the outgrowths, such as the details of child care and of kin classification, represent only slight alterations that have not yet concealed their Pleistocene origins. Others, such as religion and class structure, are such gross transmutations that only the combined resources of anthropology and history can hope to trace their cultural phylogeny back to rudiments in the hunter-gatherers' repertory. But even these might in time be subject to a statistical characterization consistent with biology.

The most extreme and significant hypertrophic segment is the gathering and sharing of knowledge. Science and technology expand at an accelerating rate in ways that alter our existence year by year. To judge realistically the magnitude of that growth, note that it is already within our reach to build computers with the memory capacity of a human brain. Such an instrument is admittedly not very practical: it would occupy most of the space of the Empire State Building and draw down an amount of energy equal to half the output of the Grand Coulee Dam. In the 1980s, however, when new "bubble memory" elements already in the experimental stage are added, the computer might be shrunk to fill a suite of offices on one floor of the same building. Meanwhile, advances in storage and retrieval are matched by increases in the rate of flow of information. During the past twenty-five years transoceanic telephone calls and amateur radio transmission have increased manyfold, television has become global, the number of books and journals has grown exponentially, and universal literacy has become the goal of most nations. The fraction of Americans working in occupations concerned primarily with information has increased from 20 to nearly 50 percent of the work force.

Pure knowledge is the ultimate emancipator. It equalizes people and sovereign states, erodes the archaic barriers of superstition and promises to lift the trajectory of cultural evolution. But I do not believe it can change the ground rules of human behavior or alter the main course of history's predictable trajectory. Self-knowledge will reveal the elements of biological human nature from which modern social life proliferated in all its strange forms. It will help to distinguish safe from dangerous future courses of action with greater precision. We can hope to decide more judiciously which of the elements of human nature to cultivate and which to subvert, which to take open pleasure with and which to handle with care. We will not, however, eliminate the hard biological substructure until such time, many years from now, when our descendents may learn to change the genes themselves. With that basic proposition having been stated, I now invite you to reconsider four of the elemental categories of behavior, aggression, sex, altruism, and religion, on the basis of sociobiological theory.

Chapter 5. Aggression

Are human beings innately aggressive? This is a favorite question of college seminars and cocktail party conversations, and one that raises emotion in political ideologues of all stripes. The answer to it is yes. Throughout history, warfare, representing only the most organized technique of aggression, has been endemic to every form of society, from hunter-gatherer bands to industrial states. During the past three centuries a majority of the countries of Europe have been engaged in war during approximately half of all the years; few have ever seen a century of continuous peace. Virtually all societies have invented elaborate sanctions against rape, extortion, and murder, while regulating their daily commerce through complex customs and laws designed to minimize the subtler but inevitable forms of conflict. Most significantly of all, the human forms of aggressive behavior are species-specific: although basically primate in form, they contain features that distinguish them from aggression in all other species. Only by redefining the words "innateness" and "aggression" to the point of uselessness might we correctly say that human aggressiveness is not innate.

Theoreticians who wish to exonerate the genes and blame human aggressiveness wholly on perversities of the environment point to the tiny minority of societies that appear to be nearly or entirely pacific. They forget that innateness refers to the measurable probability that a trait will develop in a specified set of environments, not to the certainty that the trait will develop in all environments. By this criterion human beings have a marked hereditary predisposition to aggressive behavior. In fact, the matter is even more clearcut than this qualification implies. The most peaceable tribes of today were often the ravagers of yesteryear and will probably again produce soldiers and murderers in the future. Among contemporary !Kung San violence in adults is almost unknown; Elizabeth Marshall Thomas has correctly named them the "harmless people." But as recently as fifty years ago, when these "Bushman" populations were denser and less rigidly controlled by the central government, their homicide rate per capita equalled that of Detroit and Houston. The Semai of Malaya have shown an even greater plasticity. Most of the time they seem to be innocent of even the concept of violent aggression. Murder is unknown, no explicit word for kill exists ("hit" is the preferred euphemism), children are not struck, and chickens are beheaded only as a much regretted necessity. Parents carefully train their children in these habits of nonviolence. When Semai men were recruited by the British colonial government to join in the campaign against Communist guerillas in the early 1950s, they were simply unaware that soldiers are supposed to fight and kill. "Many people who knew the Semai insisted that such an unwarlike people could never make good soldiers," writes the American anthropologist Robert K. Dentan. But they were proved wrong:

> Communist terrorists had killed the kinsmen of some of the Semai counterinsurgency troops. Taken out of their nonviolent society and ordered to kill, they seem to have been swept up in a sort of insanity which they call "blood drunkenness." A typical veteran's story runs like this. "We killed, killed,

killed. The Malays would stop and go through people's pockets and take their watches and money. We did not think of watches or money. We thought only of killing. Wah, truly we were drunk with blood." One man even told how he had drunk the blood of a man he had killed.

Like most other mammals, human beings display a behavioral scale, a spectrum of responses that appear or disappear according to particular circumstances. They differ genetically from many other animal species that lack such a pattern of behavior altogether. Because there is a complex scale instead of a simple, reflex-like response, psychoanalysts and zoologists alike have had an extraordinarily difficult time arriving at a satisfactory general characterization of human aggression. They would encounter exactly the same difficulty defining gorilla aggression or tiger aggression. Freud interpreted the behavior in human beings as the outcome of a drive that constantly seeks release. Konrad Lorenz, in his book *On Aggression*, modernized this view with new data from the studies of animal behavior. He concluded that human beings share a general instinct for aggressive behavior with other animal species. This drive must somehow be relieved, if only through competitive sports. Erich Fromm, in *The Anatomy of Human Destructiveness*, took a different and still more pessimistic view that man is subject to a unique death instinct that commonly leads to pathological forms of aggression beyond those encountered in animals.

Both of these interpretations are essentially wrong. Like so many other forms of behavior and "instinct," aggression in any given species is actually an ill-defined array of different responses with separate controls in the nervous system. No fewer than seven categories can be distinguished: the defense and conquest of territory, the assertion of dominance within well-organized groups, sexual aggression, acts of hostility by which weaning is terminated, aggression against prey, defensive counterattacks against predators, and moralistic and disciplinary aggression used to enforce the rules of society. Rattlesnakes provide an instructive example of the distinctions between these basic categories. When two males compete for access to females, they intertwine their necks and wrestle as though testing each other's strength, but they do not bite, even though their venom is as lethal to other rattlesnakes as it is to rabbits and mice. When a rattlesnake stalks its prey it strikes from any number of positions without advance warning. But when the tables are turned and the snake is confronted by an animal large enough to threaten its safety, it coils, pulls its head forward to the center of the coil in striking position, and raises and shakes its rattle. Finally, if the intruder is a king snake, a species specialized for feeding on other snakes, the rattlesnake employs a wholly different maneuver: it coils, hides its head under its body, and slaps at the king snake with one of the raised coils. So to understand the aggression of rattlesnakes or human beings it is necessary to specify which of the particular forms of aggressive behavior is of interest.

Continuing research in zoology has also established that none of the categories of aggressive behavior exists in the form of a general instinct over broad arrays of species. Each category can be added, modified, or erased by an individual species during the course of its genetic evolution, in the same way that eye color can be altered from one shade to another or a particular skin gland added or eliminated. When natural selection is intense, these changes can occur throughout an entire population in only a few generations. Aggressive behavior is in fact one of the genetically most labile of all traits. We commonly find that one species of bird or mammal is highly territorial, with every square meter of habitable environment carefully staked out; the residents perform spectacular dances or emit loud cries and noisome odors to repel rivals of the same species from their private little domains. Yet coexisting in the same habitats may be a second, otherwise similar species that shows no trace of territorial behavior. Equally abrupt differences among species commonly occur in the other categories of aggression. In short, there is no evidence that a widespread unitary aggressive instinct exists.

The reason for the absence of a general aggressive instinct has been revealed by research in ecology. Most kinds of aggressive behavior among members of the same species are responsive to crowding in the environment. Animals use aggression as a technique for gaining control over necessities, ordinarily food or shelter, that are scarce or are likely to become so at some time during the life cycle. They intensify their threats and attack with increasing frequency as the population around them grows denser. As a result the behavior itself induces members of the population to spread out in space, raises the death rate, and lowers the birth rate. In such cases aggression is said to be a "density-dependent factor" in controlling population growth. As it gradually increases in intensity, it operates like a tightening valve to slow and finally shut off the increase in numbers. Other species, in contrast, seldom or never run short of the basic necessities of life. Their numbers are reduced instead by the density-dependent effects of predators, parasites, or emigration. Such animals are typically pacific toward each other, because they rarely grow numerous enough for aggressive behavior to be of any use to individuals. And if aggression confers no advantage, it is unlikely to be encoded through natural selection into the innate behavioral repertoire of the species.

Journalists following the lead of Lorenz and Fromm have in the past depicted humankind as bloodthirsty beyond the explanatory powers of science. Yet this too is wrong. Although markedly predisposed to aggressiveness, we are far from being the most violent animal. Recent studies of hyenas, lions, and langur monkeys, to take three familiar species, have disclosed that individuals engage in lethal fighting, infanticide, and even cannibalism at a rate far above that found in human societies. When a count is made of the number of murders committed per thousand individuals per year, human beings are well down on the list of violently aggressive creatures, and I am confident that this would still be the case even if our episodic wars were to be averaged in. Hyena packs even clash in deadly pitched battles that are virtually indistinguishable from primitive human warfare. Here is an account by Hans Kruuk, a zoologist at Oxford University, of a dispute over a newly killed wildebeest:

> The two groups mixed with an uproar of calls, but within seconds the sides parted again and the Mungi hyenas ran away, briefly pursued by the Scratching Rock hyenas, who then returned to the carcass. About a dozen of the Scratching Rock hyenas, though, grabbed one of the Mungi males and bit him wherever they could — especially in the belly, the feet and the ears. The victim was completely covered by his attackers, who proceeded to maul him for about 10 minutes while their clan fellows were eating the wildebeest. The Mungi male was literally pulled apart, and when I later studied the injuries more closely, it appeared that his ears were bitten off and so were his feet and testicles, he was paralyzed by a spinal injury, had large gashes in the hind legs and belly, and subcutaneous hemorrhages all over . . . The next morning, I found a hyena eating from the carcass and saw evidence that more had been there; about one-third of the internal organs and muscles had been eaten. Cannibals!

Comparable episodes are becoming commonplace in the annals of the natural history of other kinds of mammals. I suspect that if hamadryas baboons had nuclear weapons, they would destroy the world in a week. And alongside ants, which conduct assassinations, skirmishes, and pitched battles as routine business, men are all but tranquilized pacifists. For those who wish to confirm this statement directly, ant wars are very easy to observe in most towns and cities in the eastern United States. One simply looks for masses of small blackish brown ants struggling together on sidewalks or lawns. The combatants are members of rival colonies of the common pavement ant, *Tetramorium caespitum.* Thousands of individuals may be involved, and the battlefield typically occupies several square feet of the grassroots jungle.

Finally, the more violent forms of human aggression are not the manifestations of inborn drives that periodically break through dams of inhibition. The "drive-discharge" model created by Freud and Lorenz has been replaced by a more subtle explanation based on the interaction of genetic potential and learning. The most persuasive single piece of evidence for the latter, "culture-pattern" model has been provided by Richard G. Sipes, an anthropologist. Sipes noted that if aggression is a quantity in the brain that builds up and is released, as suggested by the drive-discharge model, then it can take the form of either war or the most obvious substitutes of war, including combative sports, malevolent witchcraft, tatooing and other ritualized forms of body mutilation, and the harsh treatment of deviates. As a consequence, warlike activities should result in a reduction of its lesser substitutes. If, in contrast, violent aggression is the realization of a potential that is enhanced by learning, an increase in the practice of war should be accompanied by an increase in the substitutes. By comparing the qualities of ten notably warlike societies with those of ten pacific societies, Sipes found that the culture-pattern model is upheld over the rival drive-discharge hypothesis: the practice of war is accompanied by a greater development of combatant sports and other lesser forms of violent aggression.

The clear perception of human aggressive behavior as a structured, predictable pattern of interaction between genes and environment is consistent with evolutionary theory. It should satisfy both camps in the venerable nature-nurture controversy. On the one hand it is true that aggressive behavior, especially in its more dangerous forms of military action and criminal assault, is learned. But the learning is prepared, in the sense explained in Chapter 3; we are strongly predisposed to slide into deep, irrational hostility under certain definable conditions. With dangerous ease hostility feeds on itself and ignites runaway reactions that can swiftly progress to alienation and violence. Aggression does not resemble a fluid that continuously builds pressure against the walls of its containers, nor is it like a set of active ingredients poured into an empty vessel. It is more accurately compared to a preexisting mix of chemicals ready to be transformed by specific catalysts that are added, heated, and stirred at some later time.

The products of this neural chemistry are aggressive responses that are distinctively human. Suppose that we could enumerate all of the possible kinds of actions in all species. In this imaginary example, there might be exactly twenty-three such responses, which could be labeled A through W. Human beings do not and cannot manifest every behavior; perhaps all of the societies in the world taken together employ A through P. Furthermore, they do not develop each of the options with equal facility; there is a strong tendency under all existing conditions of child rearing for behaviors A through G to appear, and consequently H through P are encountered in very few cultures. It is the *pattern* of such probabilities that is inherited. We say that for each environment there is a corresponding probability distribution of responses. To make the statistical characterization entirely meaningful, we must then go on to compare human beings with other species. We note that rhesus monkeys can perhaps develop only aggressive behaviors F through J, with a strong bias toward F and G, while one kind of termite can show only A and another kind of termite only B. Which behavior particular human beings display depends on what they experience within their own culture, but the total array of human possibilities, like the monkey array or termite array, is inherited. It is the evolution of each pattern that sociobiologists attempt to analyze.

Territoriality is one of the variants of aggressive behavior that can be directly evaluated by the new insights of biology. Students of animal behavior define a territory as an area occupied more or less exclusively either directly by overt defense or indirectly through advertisement. This area invariably contains a scarce resource, usually a steady food supply, shelter, space for sexual display, or a site for laying eggs. Often the limitation on the availability of the resource to competing individuals secondarily affects population growth to the extent of also serving as a density-dependent factor, so that territorial defense intervenes as a buffering device against long-term changes in the environment. In other words, territoriality prevents the population from either exploding or crashing. Close studies by zoologists of the daily schedules, feeding behavior, and energy expenditures of individual animals have revealed that territorial behavior evolves in animal species only when the vital resource is *economically defensible*: the energy saved and the increase in survival and reproduction due to territorial defense outweigh the energy expended and the risk of injury and death. The researchers have been able to go further in some instances to prove that in the case of food territories the size of the defended area is at or just above the size required to yield enough food to keep the resident healthy and able to reproduce. Finally, territories contain an "invincible center." The resident animal defends the territory far more vigorously than intruders attempt to usurp it, and as a result the defender usually wins. In a special sense, it has the "moral advantage" over trespassers.

The study of territorial behavior in human beings is in a very early stage. We know that bands of hunter-gatherers around the world are commonly aggressive in their defense of land that contains a reliable food resource. The Guayaki Indians of Paraguay jealously guard their hunting grounds and regard trespassing as the equivalent of a declaration of war. Before their societies were destroyed by European influence, the Ona of Tierra del Fuego were most likely to raid neighbors who trespassed in pursuit of guanaco. Similarly, the Washo Indians of the Great Basin attacked bands who fished "their" lakes or hunted "their" deer in the more stable portions of the winter home ranges. The Nyae Nyae Bushmen believed that they had the right to kill neighbors who gathered vital plant foods from their foraging areas. The Walbiri of the Australian desert were especially concerned over water holes. One band could enter the range of another only by permission, and trespassers were likely to be killed. Early observers recorded one pitched battle among Walbiri for the control of water wells in which more than twenty tribesmen were killed on each side.

Although these anecdotes have been known for a long time, it is only very recently that anthropologists have begun to analyze the evidences of human territory with the basic theory of animal ecology. Rada Dyson-Hudson and Eric A. Smith have noted that areas defended by hunter-gatherers are precisely those that appear to be the most economically defensible. When food resources are scattered in space and unpredictable in time, the bands do not defend their home ranges and in fact often share occasional discoveries of rich food sources. The Western Shoshoni, for example, occupied an arid portion of the Great Basin in which the amount of game and most plant foods was poor and unpredictable. Their population density was very low, about one person in twenty square miles, and hunting and foraging were usually conducted by solitary individuals or families. Their home ranges were correspondingly huge, and they were forced into a nomadic existence. Families shared information on good piñon crops, concentrations of locusts, and forthcoming rabbit drives. Western Shoshoni seldom aggregated long enough to form bands or villages. They had no concept of ownership of land or any resource on it, with the single exception of eagle nests.

In contrast, the Owens Valley Paiute occupied relatively fertile land with denser stands of piñon pine and abundant game. Groups of villages were organized into bands, each of which owned sections of the valley that cut across the Owens River and extended up the mountains on either side. These territories were defended by means of social and religious sanctions reinforced with occasional threats and attacks. At most, the residents invited members of other bands, especially their relatives, to pick piñon nuts on their land.

The flexibility displayed by the Great Basin tribes parallels that occurring among other populations and species of mammals. In both men and animals its expression is correlated with the richness and spatial distribution of the most vital resources within the home range. But the range of expression is a characteristic of each species, and the total range of human beings, although unusually broad, does not encompass all of the animal patterns combined. In that sense human territorial behavior is genetically limited in its expression.

The biological formula of territorialism translates easily into the rituals of modern property ownership. When described by means of generalizations clear of emotion and fictive embellishment this behavior acquires new flavor — at once intimately familiar, because our own daily lives are controlled by it, and yet distinctive and even very peculiar, because it is after all a diagnostic trait of just one mammalian species. Each culture develops its own particular rules to safeguard personal property and space. Pierre van den Berghe, a sociologist, has provided the following description of present-day behavior around vacation residences near Seattle:

> Before entering familial territory, guests and visitors, especially if they are unexpected, regularly go through a ritual of identifi-

cation, attention drawing, greeting and apology for the possible disturbance. This behavioral exchange takes place outdoors if the owner is first encountered there, and is preferably directed at adults. Children of the owners, if encountered first, are asked about the whereabouts of their parents. When no adult owners are met outdoors, the visitor typically goes to the dwelling door, where he makes an identifying noise, either by knocking on the door or ringing a bell if the door is closed, or by voice if the door is open. The threshold is typically crossed only on recognition and invitation by the owner. Even then, the guest feels free to enter only the sitting room, and usually makes additional requests to enter other parts of the house, such as a bathroom or bedroom.

When a visitor is present, he is treated by the other members of the [vacation residence] club as an extension of his host. That is, his limited privileges of territorial occupancy extend only to the territory of his host, and the host will be held responsible by other owners for any territorial transgressions of the guests . . . Children, too, are not treated as independent agents, but as extensions of their parents or of the adult "responsible" for them, and territorial transgressions of children, especially if repeated, are taken up with the parents or guardians.

The dirt road through the development is freely accessible to all members of the club who use it both to gain access to their lots and to take walks. Etiquette calls for owners to greet each other when seeing each other outdoors, but owners do not feel free to enter each other's lots without some ritual of recognition. This ritual is, however, less formal and elaborate when entering lots outdoors than when entering houses.

War can be defined as the violent rupture of the intricate and powerful fabric of the territorial taboos observed by social groups. The force behind most warlike policies is ethnocentrism, the irrationally exaggerated allegiance of individuals to their kin and fellow tribesmen. In general, primitive men divide the world into two tangible parts, the near environment of home, local villages, kin, friends, tame animals, and witches, and the more distant universe of neighboring villages, intertribal allies, enemies, wild animals, and ghosts. This elemental topography makes easier the distinction between enemies who can be attacked and killed and friends who cannot. The contrast is heightened by reducing enemies to frightful and even subhuman status.

The Mundurucú headhunters of Brazil made all these distinctions and in addition literally turned their enemies into game. The warriors spoke of the *pariwat* (non-Mundurucú) in the same language ordinarily reserved for peccary and tapir. A high status was conferred on the taker of a human trophy head. He was believed to have attained special influence with the supernatural powers of the forest. Warfare was refined into a high art, in which other tribes were skillfully hunted as though they were packs of especially dangerous animals.

The raids were planned with great care. In the cover of the predawn darkness the Mundurucú men circled the enemy village, while their shaman quietly blew a sleep trance on the people within. The attack began at dawn. Incendiary arrows were shot onto the thatched houses, then the attackers ran screaming out of the forest into the village, chased the inhabitants into the open, and decapitated as many adult men and women as possible. Because annihilation of an entire village was difficult and risky, the attackers soon retreated with the heads of their victims. They proceeded on forced march as far as they could before resting, then headed home or on to the next enemy village.

William H. Durham, who reanalyzed Robert F. Murphy's data

on the Mundurucú, has presented a convincing case that warfare and the game metaphor are direct adaptations that benefit the individual fitness of the headhunter warriors. In the traditional manner of the natural sciences, Durham applied the evidences of Mundurucú and other primitive warfare to a set of three mutually exclusive and competing hypotheses, which in this instance appear to exhaust the possibilities of the relation between heredity and culture.

HYPOTHESIS 1: *Cultural traditions of warfare in primitive societies evolved independently of the ability of human beings to survive and reproduce.* People fight wars for various and sundry cultural reasons which have no consistent relation to genetic fitness, that is, to the survival and reproductive success of the individual and his close kin. Primitive war is not well explained by the principles of sociobiology; it is better understood as a purely cultural phenomenon, the product of social organization and political arrangements which themselves have nothing to do with fitness.

HYPOTHESIS 2: *Cultural traditions of primitive warfare evolved by selective retention of traits that increase the inclusive genetic fitness of human beings.* People fight wars when they and their closest relatives stand to gain long-term reproductive success, in competition both with other tribes and with other members of their own tribe. Despite appearances to the contrary, warfare may be just one example of the rule that cultural practices are generally adaptive in a Darwinian sense.

HYPOTHESIS 3: *Cultural traditions of primitive warfare evolved by a process of group selection that favored the self-sacrificing tendencies of some warriors.* The warriors fight battles for the good of the group and do not therefore expect net benefits for themselves and their immediate kin. The tribe that prevailed was able to expand by increasing the absolute number of its altruistic warriors, even though this genetic type declined relative to the other members of the tribe during episodes of warfare. The proneness toward violent aggression is a good example that cultural practices are directed to some extent by genetic traits favoring entire groups while disfavoring the individual members that display them.

In the case of the Mundurucú headhunters, it is the second hypothesis that best explains the actions of the warriors. Ferocity and bravery confer direct and tangible benefits on the individuals exhibiting these qualities. Although solid demographic proof is absent, indirect evidence suggests that numbers of the Mundurucú were (and still are, in a pacified state) limited by the shortage of high-quality protein. The prevailing density-dependent factor in the environment of the aboriginal savanna settlements of the Mundurucú appears to have been the quantity of game, especially peccaries, in nearby rain forests. Hunting was a major daily occupation of the men. They ordinarily worked in groups, because peccaries travel in herds, and afterward they divided the game among the families of their village in accordance with strict rules. Surrounding tribes competed for the same resource in the overlapping hunting ranges. When these competitors were decimated by murderous attacks, the Mundurucú share of the forest's yield was correspondingly increased. The biological effect of warfare on the successful Mundurucú headhunters appears to have been straightforward.

Yet the Mundurucú themselves were not directly aware of any Darwinian edge. Their justification for warlike behavior was richly overlaid by the powerful but opaque sanctions of custom and religion. Headhunting was simply a given of their existence. Neither defense of territory nor provocation by other groups was remembered as a cause of war in tribal lore. Non-Mundurucú were victims by definition. "It might be said that enemy tribes caused the Mundurucú to go to war simply by existing," Murphy writes, "and the word for enemy meant merely any group that was not Mundurucú." Traditional religious practices were centered on supplications for the abundance of game and the ritual observance of rules for its conservation. The Mundurucú believed that supernatural spirit "mothers" were poised to take swift vengeance on the hunter who killed for the hide and left the carcass to rot. So it is not very surprising that the concept of the enemy was subordinated to the concept of game. Or that the successful headhunter should be called *Dajeboisi* — "mother of the peccary." Yet the Mundurucú did not arrive at these prescriptions through understanding the ecological principles of interference competition, density dependence, and animal and human demography. They invented a simpler and more vivid universe of friends, enemies, game, and the mediating spirits of the forest that serve the same end as a scientific understanding of ecology.

The particular forms of organized violence are not inherited. No genes differentiate the practice of platform torture from pole and stake torture, headhunting from cannibalism, the duel of champions from genocide. Instead there is an innate predisposition to manufacture the cultural apparatus of aggression, in a way that separates the conscious mind from the raw biological processes that the genes encode. Culture gives a particular form to the aggression and sanctifies the uniformity of its practice by all members of the tribe.

The cultural evolution of aggression appears to be guided jointly by the following three forces: (1) genetic predisposition toward learning some form of communal aggression; (2) the necessities imposed by the environment in which the society finds itself; and (3) the previous history of the group, which biasses it toward the adoption of one cultural innovation as opposed to another. To return to the more general metaphor used in developmental biology, the society undergoing cultural evolution can be said to be moving down the slope of a very long developmental landscape. The channels of formalized aggression are deep; culture is likely to turn into one or the other but not to avoid them completely. These channels are shaped by interaction between the genetic predisposition to learn aggressive responses and the physical properties of the home range that favor particular forms of the responses. Society is influenced to take a particular direction by idiosyncratic features of its pre-existing culture.

Thus the Mundurucú populations were apparently limited by scarcity of high-grade protein, and they perfected headhunting as the convention by which competition was diminished on the hunting grounds. The Yanomamö of southern Venezuela and northern Brazil, in contrast, are temporarily in the midst of rapid population growth and range expansion. Reproduction by the men is limited not by food but by the availability of women. A principle of animal sociobiology, still only partly tested, is that in times of plenty and in the absence of effective predators females tend to become a density-dependent factor limiting population growth. As Napoleon Chagnon has shown, the Yanomamö conduct their wars over women and in order to revenge deaths that ultimately trace back to competition for women. This is not a casual or frivolous preoccupation. They have been aptly called the "fierce people." One village studied by Chagnon was raided twenty-five times in nineteen months by neighboring villages. One quarter of all Yanomamö men die in battle, but the surviving warriors are often wildly successful in the game of reproduction. The founder of one bloc of villages had forty-five children by eight wives. His sons were also prolific, so that approximately 75 percent of all of the sizable population in the village bloc were his descendants.

It is obvious that the specific conventions of aggression — for example ambush as opposed to open warfare, and ornamental stone axes as opposed to bamboo spears — are heavily influenced by the materials at hand and the bits and pieces of past custom that can be conveniently adapted. In Claude Lévi-Strauss's nice expression, culture uses the *bricolage* available to it. What is less obvious is the process that predisposes people to fabricate aggressive cultures. Only by considering the determinants of aggression at the three levels — the ultimate, biological predisposition; the requirements of the present environment; and the accidental details that contribute to cultural drift — can we fully comprehend its evolution in human societies.

Although the evidence suggests that the biological nature of humankind launched the evolution of organized aggression and roughly directed its early history across many societies, the eventual outcome of that evolution will be determined by cultural processes brought increasingly under the control of rational thought. The practice of war is a straightforward example of a hypertrophied biological predisposition. Primitive men cleaved their universe into friends and enemies and responded with quick, deep emotion to even the mildest threats emanating from outside the arbitrary boundary. With the rise of chiefdoms and states, this tendency became institutionalized, war was adopted as an instrument of policy of some of the new societies, and those that employed it best became — tragically — the most successful. The evolution of warfare was an autocatalytic reaction that could not be halted by any people, because to attempt to reverse the process unilaterally was to fall victim. A new mode of natural selection was operating at the level of entire societies. In his pioneering work on the subject Quincy Wright wrote:

> Out of the warlike peoples arose civilization, while the peaceful collectors and hunters were driven to the ends of the earth, where they are gradually being exterminated or absorbed, with only the dubious satisfaction of observing the nations which had wielded war so effectively to destroy them and to become great, now victimized by their own instrument.

Keith Otterbein, an anthropologist, has studied quantitatively the variables affecting warlike behavior in forty-six cultures, from the relatively unsophisticated Tiwi and Jivaro to more advanced societies such as the Egyptians, Aztecs, Hawaiians, and Japanese. His main conclusions will cause no great surprise: as societies become centralized and complex, they develop more sophisticated military organizations and techniques of battle, and the greater their military sophistication, the more likely they are to expand their territories and to displace competing cultures.

Civilizations have been propelled by the reciprocating thrusts of cultural evolution and organized violence, and in our time they have come to within one step of nuclear annihilation. Yet when countries have reached the brink, in the Formosan Straits, Cuba, and the Middle East, their leaders have proved able to turn back. In Abba Eban's memorable words on the occasion of the 1967 Arab-Israeli war, men use reason as a last resort.

Not only that, but the full evolution of warfare can be reversed, even in the face of entrenched cultural practice. In pre-European times the Maori of New Zealand were among the most aggressive people on earth. Raids among their forty tribes were frequent and bloody. Insults, hostility, and retribution were carefully tallied in tribal memories. Defense of personal honor and courage were the paramount virtues, victory by force of arms the highest achievement. According to Andrew Vayda, an expert on primitive war, the prime mover of Maori warfare was ecological competition. Revenge led to open fighting for land and then to territorial conquests. Alliances were based on kinship; the Maoris consciously and explicitly expanded against the territories of the genealogically most distant lineages. In 1837, when Hokianga warriors arrived at one fight already in progress between two sections of the Nga Puhi tribe, they were undecided about the side to join, because they were equally related to both. The major effect of these territorial wars was stabilization of the population. As groups became overcrowded, they expanded by displacing and reducing rival groups. The Maori population was a constantly shifting mosaic of tribal groups held at a level density overall, like the lion populations of Kenya, by territorial aggression acting as an ecological control.

This terrible equilibrium was finally disrupted and reversed when European firearms were introduced. The Maoris were understandably enchanted by the first muskets that the British colonists showed them. One traveller recorded such an encounter around 1815:

> Firing with my fowling-piece, at a bird that had settled on an adjacent tree, I happened to kill it, and this instantly threw the whole village, men, women, and children into violent confusion; who, knowing not how to account for the seeming phenomenon, testified the appalling effect it had upon them, by setting up a tremendous shout, and astounding my ears with their uproar. While in the act of shewing them the bird I had killed, which they examined very attentively, perceiving another on the same tree, I fired at this also, and brought it down; which occasioned a repetition of their amazement and made them vociferate even louder than at first.

Within a few years Maori leaders acquired guns of their own and began to employ them with devastating effect on their neighbors. One individual, the Nga Puhi chief Hongi Hiki, bought 300 guns from British traders and launched a brief career as a conqueror. Before his death in 1828, he and his allies led numerous expeditions and killed thousands of people. While their immediate motivation was revenge for old defeats, they not coincidentally extended the power and territory of the Nga Puhi. Other tribes rushed to arm themselves in order to regain parity in the escalating hostilities.

The arms race soon became self-limiting. Even the victors paid a heavy price. To obtain more muskets, the Maoris devoted inordinate amounts of their time to producing flax and other goods that could be traded to the Europeans for guns. And in order to grow

more flax many moved to the swampy lowlands, where large numbers died of disease. During the approximately twenty years of mus-
ket war, fully one quarter of the population died from one cause or
another related to the conflict. By 1830 the Nga Puhi had begun to
question the use of fighting for revenge; the old values crumbled
soon afterward. In the late 1830s and early 1840s the Maoris as a
whole converted rapidly and massively to Christianity, and warfare
among the tribes ceased entirely.

To recapitulate the total argument, human aggression cannot be
explained as either a dark-angelic flaw or a bestial instinct. Nor is it
the pathological symptom of upbringing in a cruel environment.
Human beings are strongly predisposed to respond with unreasoning
hatred to external threats and to escalate their hostility sufficiently
to overwhelm the source of the threat by a respectably wide margin
of safety. Our brains do appear to be programmed to the following
extent: we are inclined to partition other people into friends and
aliens, in the same sense that birds are inclined to learn territorial
songs and to navigate by the polar constellations. We tend to fear
deeply the actions of strangers and to solve conflict by aggression.
These learning rules are most likely to have evolved during the past
hundreds of thousands of years of human evolution and, thus, to have
conferred a biological advantage on those who conformed to them
with the greatest fidelity.

The learning rules of violent aggression are largely obsolete. We
are no longer hunter-gatherers who settle disputes with spears, arrows, and stone axes. But to acknowledge the obsolescence of the
rules is not to banish them. We can only work our way around them.
To let them rest latent and unsummoned, we must consciously un-
dertake those difficult and rarely travelled pathways in psychological development that lead to mastery over and reduction of the profound human tendency to learn violence.

The Yanomamö have been heard to say, "We are tired of fighting.

We don't want to kill anymore. But the others are treacherous and cannot be trusted." It is not hard to see that all people think the same way. With pacifism as a goal, scholars and political leaders will find it useful to deepen studies in anthropology and social psychology, and to express this technical knowledge openly as part of political science and daily diplomatic procedure. To provide a more durable foundation for peace, political and cultural ties can be promoted that create a confusion of cross-binding loyalties. Scientists, great writers, some of the more successful businessmen, and Marxist-Leninists have been doing just that more or less unconsciously for generations. If the tangle is spun still more thickly, it will become discouragingly difficult for future populations to regard each other as completely discrete on the basis of congruent distinctions in race, language, nationhood, religion, ideology, and economic interest. Undoubtedly there exist other techniques by which this aspect of human nature can be gently hobbled in the interest of human welfare.

Chapter 6. Sex

Sex is central to human biology and a protean phenomenon that permeates every aspect of our existence and takes new forms through each step in the life cycle. Its complexity and ambiguity are due to the fact that sex is not designed primarily for reproduction. Evolution has devised much more efficient ways for creatures to multiply than the complicated procedures of mating and fertilization. Bacteria simply divide in two (in many species, every twenty minutes), fungi shed immense numbers of spores, and hydras bud offspring directly from their trunks. Each fragment of a shattered sponge grows into an entire new organism. If multiplication were the only purpose of reproductive behavior, our mammalian ancestors could have evolved without sex. Every human being might be asexual and sprout new offspring from the surface cells of a neutered womb. Even now, a swift, bacterium-like method of asexual reproduction occurs on the rare occasions when identical twins are created by a single division of an already fertilized egg.

Nor is the primary function of sex the giving and receiving of pleasure. The vast majority of animal species perform the sexual act mechanically and with minimal foreplay. Pairs of bacteria and protozoans form sexual unions without the benefit of a nervous system, while corals, clams, and many other invertebrate animals simply shed their sex cells into the surrounding water — literally without giving the matter a thought, since they lack a proper brain. Pleasure is at best an enabling device for animals that copulate, a means for inducing creatures with versatile nervous systems to make the heavy investment of time and energy required for courtship, sexual intercourse, and parenting.

Moreover, sex is in every sense a gratuitously consuming and risky activity. The reproductive organs of human beings are anatomically complex in ways that make them subject to lethal malfunctions, such as ectopic pregnancy and venereal disease. Courtship activities are prolonged beyond the minimal needs of signaling. They are energetically expensive and even dangerous, to the degree that the more ardent are put at greater risk of being killed by rivals or predators. At the microscopic level, the genetic devices by which sex is determined are finely tuned and easily disturbed. In human beings one sex chromosome too few or too many, or a subtle shift in the hormone balance of a developing fetus, creates abnormalities in physiology and behavior.

Thus sex by itself lends no straightforward Darwinian advantage. Moreover, sexual reproduction automatically imposes a genetic deficit. If an organism multiplies without sex, all of its offspring will be identical to itself. If, on the other hand, an organism accepts sexual partnership with another, unrelated individual, half the genes in each of its offspring will be of alien origin. With each generation thereafter, the investment in genes per descendant will be cut in half.

So there are good reasons for reproduction to be nonsexual: It can be made private, direct, safe, energetically cheap, and selfish. Why, then, has sex evolved?

The principal answer is that sex creates diversity. And diversity is the way a parent hedges its bets against an unpredictably changing environment. Imagine a case of two animal species, both of which consist entirely of individuals carrying two genes. Let us arbitrarily label one gene *A* and the other *a*. For instance, these genes might be for brown (*A*) versus blue (*a*) eye color, or right-handedness (*A*) versus left-handedness (*a*). Each individual is *Aa* because it possesses both genes. Suppose that one species reproduces without sex. Then all the offspring of every parent will be *Aa*.

The other population uses sex for reproduction; it produces sex cells, each of which contains only one of the genes, *A* or *a*. When two individuals mate they combine their sex cells, and since each adult contributes sex cells bearing either *A* or *a*, three kinds of offspring are possible: *AA*, *Aa*, and *aa*. So, from a starting population of *Aa* individuals, asexual parents can produce only *Aa* offspring, while sexual parents can produce *AA*, *Aa*, and *aa* offspring. Now let the environment change — say a hard winter, a flood, or the invasion of a dangerous predator — so that *aa* individuals are favored. In the next generation, the sexually reproducing population will have the advantage and will consist predominantly of *aa* organisms until conditions change to favor, perhaps, *AA* or *Aa* individuals.

Diversity, and thus adaptability, explains why so many kinds of organisms bother with sexual-reproduction. They vastly outnumber the species that rely on the direct and simple but, in the long run, less prudent modes of sexless multiplication.

Then why are there usually just two sexes? It is theoretically possible to evolve a sexual system based on one sex — anatomically uniform individuals who produce identically shaped reproductive cells and combine them indiscriminately. Some lower plants do just that. It is also possible to have hundreds of sexes, which is the mode among some fungi. But a two-sex system prevails through most of the living world. This system appears to permit the most efficient possible division of labor.

The quintessential female is an individual specialized for making eggs. The large size of the egg enables it to resist drying, to survive adverse periods by consuming stored yolk, to be moved to safety by the parent, and to divide at least a few times after fertilization before needing to ingest nutrients from the outside. The male is defined as the manufacturer of the sperm, the little gamete. A sperm is a minimum cellular unit, stripped down to a head packed with DNA and powered by a tail containing just enough stored energy to carry the vehicle to the egg.

When the two gametes unite in fertilization they create an instant mixture of genes surrounded by the durable housing of the egg. By cooperating to create zygotes, the female and male make it more likely that at least some of their offspring will survive in the event of a changing environment. A fertilized egg differs from an asexually reproducing cell in one fundamental respect: it contains a newly assembled mixture of genes.

The anatomical difference between the two kinds of sex cell is often extreme. In particular, the human egg is eighty-five thousand times larger than the human sperm. The consequences of this gametic dimorphism ramify throughout the biology and psychology of human sex. The most important immediate result is that the female places a greater investment in each of her sex cells. A woman can expect to produce only about four hundred eggs in her lifetime. Of these a maximum of about twenty can be converted into healthy infants. The costs of bringing an infant to term and caring for it afterward are relatively enormous. In contrast, a man releases 100 million sperm with each ejaculation. Once he has achieved fertilization his purely physical commitment has ended. His genes will benefit equally with those of the female, but his investment will be far less than hers unless she can induce him to contribute to the care of the offspring. If a man were given total freedom to act, he could theoretically inseminate thousands of women in his lifetime.

The resulting conflict of interest between the sexes is a property of not only human beings but also the majority of animal species.

Males are characteristically aggressive, especially toward one another and most intensely during the breeding season. In most species, assertiveness is the most profitable male strategy. During the full period of time it takes to bring a fetus to term, from the fertilization of the egg to the birth of the infant, one male can fertilize many females but a female can be fertilized by only one male. Thus if males are able to court one female after another, some will be big winners and others will be absolute losers, while virtually all healthy females will succeed in being fertilized. It pays males to be aggressive, hasty, fickle, and undiscriminating. In theory it is more profitable for females to be coy, to hold back until they can identify males with the best genes. In species that rear young, it is also important for the females to select males who are more likely to stay with them after insemination.

Human beings obey this biological principle faithfully. It is true that the thousands of existing societies are enormously variable in the details of their sexual mores and the division of labor between the sexes. This variation is based on culture. Societies mold their customs to the requirements of the environment and in so doing duplicate in totality a large fraction of the arrangements encountered throughout the remainder of the animal kingdom: from strict monogamy to extreme forms of polygamy, and from a close approach to unisex to extreme differences between men and women in behavior and dress. People change their attitudes consciously and at will; the reigning fashion of a society can shift within a generation. Nevertheless, this flexibility is not endless, and beneath it all lie general features that conform closely to the expectations from evolutionary theory. So let us concentrate initially on the biologically significant generalities and defer, for the moment, consideration of the undeniably important plasticity controlled by culture.

We are, first of all, moderately polygynous, with males initiating most of the changes in sexual partnership. About three-fourths of all human societies permit the taking of multiple wives, and most of them encourage the practice by law and custom. In contrast, marriage to multiple husbands is sanctioned in less than one percent of societies. The remaining monogamous societies usually fit that category in a legal sense only, with concubinage and other extramarital strategems being added to allow de facto polygyny.

Because women are commonly treated by men as a limiting resource and hence as valued property, they are the beneficiaries of hypergamy, the practice of marrying upward in social position. Polygyny and hypergamy are essentially complementary strategies. In diverse cultures men pursue and acquire, while women are protected and bartered. Sons sow wild oats and daughters risk being ruined. When sex is sold, men are usually the buyers. It is to be expected that prostitutes are the despised members of society; they have abandoned their valuable reproductive investment to strangers. In the twelfth century, Maimonides neatly expressed this biological logic as follows:

> For fraternal sentiments and mutual love and mutual help can be found in their perfect form only among those who are related by their ancestry. Accordingly a single tribe that is united through a common ancestor — even if he is remote — because of this, love one another, help one another, and have pity on one another; and the attainment of these things is the greatest purpose of the Law. Hence *harlots* are prohibited, because through them lines of ancestry are destroyed. For a child born of them is a stranger to the people; no one knows to what family group he belongs, and no one in his family group knows him; and this is the worst of conditions for him and his father.

Anatomy bears the imprint of the sexual division of labor. Men are on the average 20 to 30 percent heavier than women. Pound for pound, they are stronger and quicker in most categories of sport. The proportion of their limbs, their skeletal torsion, and the density of their muscles are particularly suited for running and throwing, the archaic specialties of the ancestral hunter-gatherer males. The world track records reflect the disparity. Male champions are always between 5 and 20 percent faster than women champions: in 1974 the difference was 8 percent in the 100 meters, 11 percent in the 400 meters, 15 percent in the mile, 10 percent in the 10,000 meters, and so on through every distance. Even in the marathon, where size and brute strength count least, the difference was 13 percent. Women marathoners have comparable endurance, but men are faster — their champions run twenty-six five-minute miles one after another. The gap cannot be attributed to a lack of incentive and training. The great women runners of East Germany and the Soviet Union are the products of nationwide recruitment and scientifically planned training programs. Yet their champions, who consistently set Olympic and world records, could not place in an average men's regional track meet. The overlap in performances between all men and women is of course great; the best women athletes are better than most male athletes, and women's track and field is an exciting competitive world of its own. But there is a substantial difference between average and best performances. The leading woman marathon runner in the United States in 1975, for example, would have ranked 752d in the national men's listing. Size is not the determinant. The smaller male runners, at 125 to 130 pounds, perform as well relative to women as do their taller and heavier competitors.

It is of equal importance that women match or surpass men in a few other sports, and these are among the ones furthest removed from the primitive techniques of hunting and aggression: long-distance swimming, the more acrobatic events of gymnastics, precision (but not distance) archery, and small-bore rifle shooting. As sports and sport-like activities evolve into more sophisticated channels dependent on skill and agility, the overall achievements of men and women can be expected to converge more closely.

The average temperamental differences between the human sexes are also consistent with the generalities of mammalian biology. Women as a group are less assertive and physically aggressive. The magnitude of the distinction depends on the culture. It ranges from a tenuous, merely statistical difference in egalitarian settings to the virtual enslavement of women in some extreme polygynous societies. But the variation in degree is not nearly so important as the fact that women differ consistently in this qualitative manner regardless of the degree. The fundamental average difference in personality traits is seldom if ever transposed.

The physical and temperamental differences between men and women have been amplified by culture into universal male dominance. History records not a single society in which women have controlled the political and economic lives of men. Even when queens and empresses ruled, their intermediaries remained primarily male. At the present writing not a single country has a woman as head of state, although Golda Meir of Israel and Indira Gandhi of India were, until recently, assertive, charismatic leaders of their countries. In about 75 percent of societies studied by anthropologists, the bride is expected to move from the location of her own family to that of her husband, while only 10 percent require the reverse exchange. Lineage is reckoned exclusively through the male line at least five times more frequently than it is through the female line. Men have traditionally assumed the positions of chieftains, shamans, judges, and warriors. Their modern technocratic counterparts rule the industrial states and head the corporations and churches.

These differences are a simple matter of record — but what is their significance for the future? How easily can they be altered?

It is obviously of vital social importance to try to make a value-free assessment of the relative contributions of heredity and environment to the differentiation of behavioral roles between the sexes. Here is what I believe the evidence shows: modest genetic differences exist between the sexes; the behavioral genes interact with virtually all existing environments to create a noticeable divergence in early psychological development; and the divergence is almost always widened in later psychological development by cultural sanctions and training. Societies can probably cancel the modest genetic differences entirely by careful planning and training, but the convergence will require a conscious decision based on fuller and more exact knowledge than is now available.

The evidence for a genetic difference in behavior is varied and substantial. In general, girls are predisposed to be more intimately sociable and less physically venturesome. From the time of birth, for example, they smile more than boys. This trait may be especially revealing, since as I showed earlier the infant smile, of all human behaviors, is most fully innate in that its form and function are virtually invariant. Several independent studies have shown that newborn females respond more frequently than males with eyes-closed, reflexive smiling. The habit is soon replaced by deliberate, communicative smiling that persists into the second year of life. Frequent smiling then becomes one of the more persistent of female traits and endures through adolescence and maturity. By the age of six months, girls also pay closer attention to sights and sounds used in communication than they do to nonsocial stimuli. Boys of the same age make no such distinction. The ontogeny then proceeds as follows: one-year-old girls react with greater fright and inhibition to clay faces, and they are more reluctant to leave their mothers' sides in novel situations. Older girls remain more affiliative and less physically venturesome than boys of the same age.

In her study of the !Kung San, Patricia Draper found no difference in the way young boys and girls are reared. All are supervised closely but unobtrusively and are seldom given any work. Yet boys wander out of view and earshot more frequently than girls, and older boys appear to be slightly more prone to join the men hunters than are girls to join the women gatherers. In still closer studies, N. G. Blurton Jones and Melvin J. Konner found that boys also engage more frequently in rough-and-tumble play and overt aggression. They also associate less with adults than do girls. From these subtle differences the characteristic strong sexual division of labor in !Kung encampments emerges by small steps.

In Western cultures boys are also more venturesome than girls and more physically aggressive on the average. Eleanor Maccoby and Carol Jacklin, in their review *The Psychology of Sex Differences*, concluded that this male trait is deeply rooted and could have a genetic origin. From the earliest moments of social play, at age 2 to 2-1/2 years, boys are more aggressive in both words and actions. They have a larger number of hostile fantasies and engage more often in mock fighting, overt threats, and physical attacks, which are directed preferentially at other boys during efforts to acquire dominance status. Other studies, summarized by Ronald P. Rohner, indicate that the differences exist in many cultures.

The skeptic favoring a totally environmental explanation might still argue that the early divergence in role playing has no biological component but is merely a response to biased training practices during very early childhood. If it occurs, the training would have to be subtle, at least partly unconscious in application, and practiced by parents around the world. The hypothesis of total environmentalism is made more improbable by recent evidence concerning the biology of hermaphrodites, who are genetically female but acquire varying degrees of masculine anatomy during the early stages of fetal development. The anomaly occurs in one of two ways.

The first is a rare hereditary condition caused by a change in a single gene site and known as the female adrenogenital syndrome.

In either sex, possession of two of the altered genes — hence, a complete lack of the normal gene in each cell of the body — prevents the adrenal glands from manufacturing their proper hormone, cortisol. In its place the adrenal glands secrete a precursor substance which has an action similar to that of the male sex hormone. If the individual is genetically male, the hormonal boost has no significant effect on sexual development. If the fetus is female, the abnormal level of male hormone alters the external genitalia in the direction of maleness. Sometimes the clitoris of such an individual is enlarged to resemble a small penis, and the labia majora are closed. In extreme cases a full penis and empty scrotum are developed.

The second means of producing the effect is by artificial hormone treatment. During the 1950s women were often given progestins, a class of artificial substances that act like progesterone, the normal hormone of pregnancy, to help them prevent miscarriages. It was discovered that in a few cases progestins, by exerting a masculinizing effect on female fetuses, transformed them into hermaphrodites of the same kind caused by the female adrenogenital syndrome.

By sheer accident the hormone-induced hermaphrodites approach a properly controlled scientific experiment designed to estimate the influence of heredity on sex differences. The experiment is not perfect, but it is as good as any other we are likely to encounter. The hermaphrodites are genetically female, and their internal sexual organs are fully female. In most of the cases studied in the United States, the external genitalia were altered surgically to an entirely female condition during infancy, and the individuals were then reared as girls. These children were subjected during fetal development to male hormones or to substances that mimic them but then "trained" to be ordinary girls until maturity. In such cases it is possible to dissect the effects of learning from the effects of deeper biological alterations, which in some cases stem directly from a known gene mutation. Behavioral maleness would almost certainly have to be ascribed to the effect of the hormones on development of the brain.

Did the girls show behavioral changes connected with their hormonal and anatomical masculinization? As John Money and Anke Ehrhardt discovered, the changes were both quite marked and correlated with the physical changes. Compared with unaffected girls of otherwise similar social backgrounds, the hormonally altered girls were more commonly regarded as tomboys while they were growing up. They had a greater interest in athletic skills, were readier to play with boys, preferred slacks to dresses and toy guns to dolls. The group with the adrenogenital syndrome was more likely to show dissatisfaction with being assigned to a female role. The evaluation of this latter group is flawed by the fact that cortisone had to be administered to the girls to offset their genetic defect. It is possible that hormone treatment alone could somehow have biased the girls toward masculine behavior. If the effect occurred it was still biological in nature, although not as deep as fetal masculinization. And of course, the effect could not have occurred in the progestin-altered girls.

So at birth the twig is already bent a little bit — what are we to make of that? It suggests that the universal existence of sexual division of labor is not entirely an accident of cultural evolution. But it also supports the conventional view that the enormous variation among societies in the degree of that division is due to cultural evolution. Demonstrating a slight biological component delineates the options that future societies may consciously select. Here the second dilemma of human nature presents itself. In full recognition of the struggle for women's rights that is now spreading throughout the world, each society must make one or the other of the three following choices:

Condition its members so as to exaggerate sexual differences in behavior. This is the pattern in almost all cultures. It results more often than not in domination of women by men and exclusion of women from many professions and activities. But this need not be the case. In theory at least, a carefully designed society with strong sexual divisions could be richer in spirit, more diversified, and even more productive than a unisex society. Such a society might safeguard human rights even while channeling men and women into different occupations. Still, some amount of social injustice would be inevitable, and it could easily expand to disastrous proportions.

Train its members so as to eliminate all sexual differences in behavior. By the use of quotas and sex-biased education it should be possible to create a society in which men and women *as groups* share equally in all professions, cultural activities, and even, to take the absurd extreme, athletic competition. Although the early predispositions that characterize sex would have to be blunted, the biological differences are not so large as to make the undertaking impossible. Such control would offer the great advantage of eliminating even the hint of group prejudice (in addition to individual prejudice) based on sex. It could result in a much more harmonious and productive society. Yet the amount of regulation required would certainly place some personal freedoms in jeopardy, and at least a few individuals would not be allowed to reach their full potential.

Provide equal opportunities and access but take no further action. To make no choice at all is of course the third choice open to all cultures. Laissez-faire on first thought might seem to be the course most congenial to personal liberty and development, but this is not necessarily true. Even with identical education for men and women and equal access to all professions, men are likely to maintain disproportionate representation in political life, business, and science. Many would fail to participate fully in the equally important, formative aspects of child rearing. The result might be legitimately viewed as restrictive on the complete emotional development of individuals. Just such a divergence and restriction has occurred in the Israeli kibbutzim, which represent one of the most powerful experiments in egalitarianism conducted in modern times.

From the time of the greatest upsurge of the kibbutz movement, in the 1940s and 1950s, its leaders promoted a policy of complete sexual equality, of encouraging women to enter roles previously reserved for men. In the early years it almost worked. The first generation of women were ideologically committed, and they shifted in large numbers to politics, management, and labor. But they and their daughters have regressed somewhat toward traditional roles, despite being trained from birth in the new culture. Furthermore, the daughters have gone further than the mothers. They now demand and receive a longer period of time each day with their children, time significantly entitled "the hour of love." Some of the most gifted have resisted recruitment into the higher levels of commercial and political leadership, so that the representation in these roles is far below that enjoyed by the same generation of men. It has been argued that this reversion merely represents the influence of the strong patriarchal tradition that persists in the remainder of Israeli society, even though the role division is now greater inside the kibbutzim than outside. The Israeli experience shows how difficult it is to predict the consequences and assess the meaning of changes in behavior based on either heredity or ideology.

From this troubling ambiguity concerning sex roles one firm conclusion can be drawn: the evidences of biological constraint alone cannot prescribe an ideal course of action. However, they can help us to define the options and to assess the price of each. The price is to be measured in the added energy required for education and reinforcement and in the attrition of individual freedom and potential. And let us face the real issue squarely: since every option has a cost, and concrete ethical principles will rarely find universal acceptance, the choice cannot be made easily. In such cases we could do well to consider the wise counsel of Hans Morgenthau: "In the combination of political wisdom, moral courage and moral judgment, man reconciles his political nature with his moral destiny. That this conciliation is nothing more than a *modus vivendi*, uneasy, precarious, and even paradoxical, can disappoint only those who prefer to gloss over and to distort the tragic contradictions of human existence with the soothing logic of a specious concord." I am suggesting that the contradictions are rooted in the surviving relics of our prior genetic history, and that one of the most inconvenient and senseless, but nevertheless unavoidable of these residues is the modest predisposition toward sex role differences.

Another residue to be weighed and measured in biological social theory is the family. The nuclear family, based on long-term sexual bonding, geographical mobility, and female domesticity, is declining at this moment in the United States. Between 1967 and 1977 the divorce rate doubled, and the number of households headed by women increased by a third. In 1977 one out of every three school children lived in a home headed by only one parent or relative, and more than half of all mothers with school-age children worked outside the home. Day care centers have come to replace the parents in many working families; their older offspring constitute a large population of "latchkey" children who are wholly unsupervised in the period between the end of school and the parents' return from work. The American birth rate has declined precipitously, from 3.80 per family in 1957 to 2.04 in 1977. Such social change in the most technologically advanced country, when correlated with the liberation of women and their massive entrance into the work force, is an event certain to have profound long-range consequences. But does it also mean that the family is a cultural artifact destined for extinction?

I think not. The family, defined broadly as a set of closely related adults with their children, remains one of the universals of human social organization. Even the societies that seem to break the rule, the Nāyar of India and the Israeli kibbutzniks, are not really autonomous social groups but special subgroups that live within larger communities. The family, taking either a nuclear or extended form, has rebounded from countless episodes of stress in many societies throughout history. In the United States, slave families were frequently broken up during sales. African customs were disregarded or discouraged, and neither marriage nor parenthood were given legal protection. Yet kin groups survived for generations, individual kin were classified, children were assigned familial surnames, and incest taboos were observed faithfully. The Africans' attachment to their families remained deep and emotional. In witness are many fragments of oral traditions and written records, such as the following letter sent by the field hand Cash and his family in 1857 after they had been separated from their closest relatives on a Georgia plantation:

> Clairssa your affectionate Mother and Father sends a heap of Love to you and your husband and my Grand Children Phebea. Mag. & Cloe. John. Judy. Sue. My aunt Aufy sinena and Minton and Little Plaska. Charles Nega. Fillis and all of their Children. Cash. Prime. Laffatte. Give our Love to Cashes brother Porter and his wife Patience. Victoria gives her Love to her Cousin Beck and Miley

According to the historian Herbert G. Gutman, networks of this kind, many unknown to the slave owners, extended throughout the South. Today, they persist with little or no dilution in the most impoverished ghettos. As Carol Stack has shown in her remarkable book *All Our Kin*, detailed knowledge of relatives and an unquestioning code of mutual loyalty are the very basis of survival among the poorest American blacks.

In some of the American communes of the 1960s and 1970s attempts were made, mostly by middle class whites, to organize themselves into egalitarian societies while rearing their children in crèches.

But, as Jerome Cohen and his coworkers have discovered, the traditional nuclear family repeatedly reasserts itself. In the end, the commune mothers expressed a need to care for their own children even stronger than that shown by mothers in ordinary married households. A third of them switched from collective parental care to the two-parent arrangement. In more traditional communities an increasing number of couples have chosen to live out of wedlock and to postpone having children. Nevertheless, the forms of their social life still resemble the classical marriage bond, and many eventually go on to raise children by conventional methods.

The human predisposition to assemble into families asserts itself even in some abnormal circumstances. At the Federal Reformatory for Women, in Alderson, West Virginia, Rose Giallombardo has found that inmates organize themselves into family-like units centered on a sexually active pair called the husband and wife. Women classified as brothers and sisters are typically added, and older inmates serve as surrogates for mothers, fathers, aunts, uncles, and even grandmothers. The roles assigned these categories parallel those found in the outside, heterosexual world. The prison pseudofamily provides its members with stability, protection, and advice, as well as food and drugs during punishment regimens. Interestingly enough, the inmates of men's prisons are organized more loosely into institution-wide hierarchies and castes, in which dominance and rank are paramount. Sexual relationships are quite common among these men, but the more passive partners, who play the female role, are ordinarily treated with contempt.

The most distinctive feature of the sexual bond, one of overriding significance for human social organization, is that it transcends sexual activity. Genetic diversification, the ultimate function of sex, is served by the physical pleasure of the sex act and outranks in importance the process of reproduction. The sexual bond is also served by pleasure, and it fulfills other roles in turn, some of which are only remotely connected to reproduction. These multiple functions and complex chains of causation are the deeper reason why sexual awareness permeates so much of human existence.

Polygyny and sexual differences in temperament can be predicted by a straightforward deduction from the general theory of evolution. But that is not possible for the covert functions of the sexual bond and the family. It is necessary to consider in addition the case histories of other species related to our own and to make ad hoc inferences concerning the actual courses of evolution. A few other primates, marmosets and gibbons in particular, have superficially human-like family groupings. Pairs of adults mate for life and cooperate to rear offspring all the way to maturity. Zoologists believe that the special forest environments in which these species live confer a Darwinian advantage on sexual bonding and family stability. They speculate that the human family also originated as an adaptation to peculiar environmental conditions, but this prevailing hypothesis is based on very few facts.

We know in particular that the earliest true men, at least back to *Homo habilis*, two to three million years ago, differed from other primates in two respects: they ranged away from the forest habitats of their ancestors, and they hunted game. The animals they captured included antelopes, elephants, and other large mammals not exploited by the mostly vegetarian monkeys and apes. These slender little people, the size of modern twelve-year-olds, were devoid of fangs and claws and almost certainly slower on foot than the four-legged animals around them. They could have succeeded in their new way of life only by relying on tools and sophisticated cooperative behavior.

What form did the new cooperation take? It might have entailed the joint and equal effort of all members of the society — men, women, and juveniles. But it could well have been based on some division of labor. Perhaps women hunted while men remained in the encampments, or the reverse, or the hunters might have been individuals above a certain size regardless of sex. In its present rudimentary state, sociobiological theory cannot predict which of these and other conceivable possibilities is the most likely. Nor is the archeological evidence from two million years ago adequate to show which one was actually used. Instead, we must rely on data from the living hunter-gatherer societies, which in their economies and population structure are closest to the ancestral human beings. Here the evidence is suggestive but not decisive.

In virtually all of the more than one hundred such societies that have been studied around the world, men are responsible for most or all of the hunting and women for most or all of the gathering. Men form organized, mobile groups that range far from the campsites in search of larger game. Women participate in the capture of smaller animals, and they collect most of the vegetable food. Although men bring home the highest grade of protein, women generally provide most of the calories. They are also frequently but not invariably responsible for the fabrication of clothing and the building of shelters.

Human beings, as typical large primates, breed slowly. Mothers carry fetuses for nine months and afterward are encumbered by infants and small children who require milk at frequent intervals through the day. It is to the advantage of each woman of the hunter-gatherer band to secure the allegiance of men who will contribute meat and hides while sharing the labor of child-rearing. It is to the reciprocal advantage of each man to obtain exclusive sexual rights to women and to monopolize their economic productivity. If the evidence from hunter-gatherer life has been correctly interpreted, the exchange has resulted in near universality of the pair bond and the prevalence of extended families with men and their wives forming the nucleus. Sexual love and the emotional satisfaction of family life can be reasonably postulated to be based on enabling mechanisms in the physiology of the brain that have been programmed to some extent through the genetic hardening of this compromise. And because men can breed at shorter intervals than women, the pair bond has been attenuated somewhat by the common practice of polygyny, the taking of multiple wives.

Human beings are unique among the primates in the intensity and variety of their sexual activity. Among other higher mammals they are exceeded in sexual athleticism only by lions. The external genitalia of both men and women are exceptionally large and advertised by tufts of pubic hair. The breasts of women are enlarged beyond the size required to house the mammary glands, while the nipples are erotically sensitive and encircled by conspicuously colored areolas. In both sexes the ear lobes are fleshy and sensitive to the touch.

Women are extraordinary in lacking the estrus, or period of heat. The females of most other primate species become sexually active, to the point of aggressiveness, only at the time of ovulation. Their genitals even swell and change color. A change in odor is probably also a general occurrence; female rhesus monkeys produce quantities of fatty acids that attract and excite the males. None of this happens in women. Their ovulation is hidden, to such a degree that it is difficult to initiate pregnancies or to avoid them even when the time of insemination is carefully selected. Women remain sexually receptive, with little variation in the capacity to respond, throughout the menstrual cycle. They never attain the peak of readiness that defines the estrus in other mammals. In the course of evolution they have eliminated the estrus by diffusing it evenly through time.

Why has sexual responsiveness become nearly continuous? The most plausible explanation is that the trait facilitates bonding; the physiological adaptation conferred a Darwinian advantage by more tightly joining the members of primitive human clans. Unusually frequent sexual activity between males and females served as the principal device for cementing the pair bond. It also reduced aggression among the males. In baboon troops and other nonhuman primate societies male hostility is intensified when females come into heat. The erasure of estrus in early human beings reduced the potential for such competition and safeguarded the alliances of hunter males.

Human beings are connoisseurs of sexual pleasure. They indulge themselves by casual inspection of potential partners, by fantasy, poetry, and song, and in every delightful nuance of flirtation leading to foreplay and coition. This has little if anything to do with reproduction. It has everything to do with bonding. If insemination were the sole biological function of sex, it could be achieved far more economically in a few seconds of mounting and insertion. Indeed, the least social of mammals mate with scarcely more ceremony. The species that have evolved long-term bonds are also, by and large, the ones that rely on elaborate courtship rituals. It is consistent with this trend that most of the pleasures of human sex constitute primary reinforcers to facilitate bonding. Love and sex do indeed go together.

The biological significance of sex has been misinterpreted by the theoreticians of Judaism and Christianity. To this day the Roman Catholic Church asserts that the primary role of sexual behavior is the insemination of wives by husbands. In his 1968 encyclical *Humanae Vitae*, which was reaffirmed by a mandate from the Congregation for the Doctrine of the Faith in 1976, Pope Paul VI prohibits the use of any form of birth control except abstinence at ovulation. Also condemned are all "genital acts" outside the framework of marriage. Masturbation is not a normal part of erotic development; it is an "intrinsically and seriously disordered act."

The Church takes its authority from natural-law theory, which is based on the idea that immutable mandates are placed by God in human nature. This theory is in error. The laws it addresses are biological, were written by natural selection, require little if any enforcement by religious or secular authorities, and have been erroneously interpreted by theologians writing in ignorance of biology. All that we can surmise of humankind's genetic history argues for a more liberal sexual morality, in which sexual practices are to be regarded first as bonding devices and only second as means for procreation.

Nowhere has the sanctification of premature biological hypothesis inflicted more pain than in the treatment of homosexuals. The Church forbids homosexual behavior. It is "intrinsically disordered." Various other cultures have agreed. At Sachsenhausen, Buchenwald, and other Nazi death camps, homosexuals wore pink triangles to distinguish them from Jews (yellow stars) and political prisoners (red triangles); later, when labor became scarce, surgeons tried to rehabilitate homosexuals by castrating them. The People's Republic of China and some other revolutionary socialist countries, fearing the deeper political implications of deviance, suppress homosexuality pro forma. In parts of the United States homophiles are still denied some of their civil liberties, while a majority of psychiatrists continue to treat homosexuality as a form of illness and express professional discouragement over its intractability.

That the moral sentinels of Western culture have condemned homosexuals is understandable. Judeo-Christian morality is based on the Old Testament, written by the prophets of an aggressive pastoral nation whose success was based on rapid and orderly population growth enhanced by repeated episodes of territorial conquest. The prescriptions of Leviticus are tailored to this specialized existence. They include the following: "You shall not lie with a man as with a woman: that is an abomination." This biblical logic seems consistent with a simplistic view of natural law when population growth is at a premium, since the overriding purpose of sexual behavior under such circumstances will seem to be the procreation of children. Most Americans still follow the archaic prescription, even though their demographic goals are now entirely different from those of the early Israelites. Homosexuals must be fundamentally deviant, the reasoning goes, because their behavior does not produce children.

There have always been a great many sinners by this definition. A generation ago Alfred Kinsey found that as many as 2 percent of American women and 4 percent of men were exclusively homosexual, while 13 percent of the men were predominantly homosexual for at least three years of their lives. Today the number of exclusive homosexuals is conservatively estimated to be five million, while gays themselves believe that the number of closet homosexuals could raise the number to twenty million. They form a consequential American subculture, employing an argot of hundreds of words and expressions. Homosexual behavior of one form or another is also common in virtually all other cultures, and in some of the high civilizations it has been permitted or approved: in classical Athenian, Persian, and Islamic societies, for example, and in late republican and early imperial Rome, in the urban, Hellenistic cultures of the Middle East, in the Ottoman Empire, and in feudal and early modern Japan.

There is, I wish to suggest, a strong possibility that homosexuality is normal in a biological sense, that it is a distinctive beneficent behavior that evolved as an important element of early human social organization. Homosexuals may be the genetic carriers of some of mankind's rare altruistic impulses.

The support for this radical hypothesis comes from certain facts considered in the new light of sociobiological theory. Homosexual behavior is common in other animals, from insects to mammals, but finds its fullest expression as an alternative to heterosexuality in the most intelligent primates, including rhesus macaques, baboons, and chimpanzees. In these animals the behavior is a manifestation of true bisexuality latent within the brain. Males are capable of adopting a full female posture and of being mounted by other males, while females occasionally mount other females.

Human beings are different in one important respect. There is a potential for bisexuality in the brain and it is sometimes expressed fully by persons who switch back and forth in their sexual preference. But in full homosexuality, as in full heterosexuality, both that choice and the symmetry of the animal pattern are lost. The preference is truly homophile: most completely homosexual men prefer masculine partners, while their female counterparts are attracted by feminine ones. As a rule, effeminate mannerisms in men are mostly unrelated to their choice of sexual partners. In modern societies, but not primitive ones, transvestites are only rarely homosexual, and the great majority of homosexual men do not differ significantly in dress and mannerisms from heterosexual men. A parallel statement can be made regarding homosexual women.

This special homophile property may hold the key to the biological significance of human homosexuality. Homosexuality is above all a form of bonding. It is consistent with the greater part of heterosexual behavior as a device that cements relationships. The predisposition to be a homophile could have a genetic basis, and the genes might have spread in the early hunter-gatherer societies because of the advantage they conveyed to those who carried them. This brings us to the nub of the difficulty, the problem most persons have in regarding homosexuality to be in any way "natural."

How can genes predisposing their carriers toward homosexuality spread through the population if homosexuals have no children? One answer is that their close relatives could have had more children as a result of their presence. The homosexual members of primitive societies could have helped members of the same sex, either while hunting and gathering or in more domestic occupations at the dwelling sites. Freed from the special obligations of parental duties, they would have been in a position to operate with special efficiency in assisting close relatives. They might further have taken the roles of seers, shamans, artists, and keepers of tribal knowledge. If the relatives — sisters, brothers, nieces, nephews, and others — were benefitted by higher survival and reproduction rates, the genes these individuals shared with the homosexual specialists would have increased at the expense of alternative genes. Inevitably, some of these genes would have been those that predisposed individuals toward homosexuality. A minority of the population would consequently always have the potential for developing homophilic preferences. Thus it is possible for homosexual genes to proliferate through collateral lines of descent, even if the homosexuals themselves do not have children. This conception can be called the "kin-selection hypothesis" of the origin of homosexuality.

The kin-selection hypothesis would be substantially supported if some amount of predisposition to homosexuality were shown to be inherited. And some evidence of such heritability does exist. Monozygotic twins, which originate from a single fertilized egg and hence are genetically identical, are more similar in the extent to which they express heterosexual or homosexual behavior than is the case for fraternal twins, which originate from separate fertilized eggs. The data, reviewed and analyzed by L. L. Heston and James Shields, suffer from the usual defects that render most twin analyses less than conclusive, but they are suggestive enough to justify further study. Some of the identical twins, according to Heston and Shields, "were not only concordant for homosexuality, but the members of each pair had developed modes of sexual behavior strikingly similar to each other. Furthermore, they did this while ignorant of their cotwin's homosexuality and, for [one pair], while widely separated geographically." Like many other human traits more confidently known to be under genetic influence, the hereditary predisposition toward homosexuality need not be absolute. Its expression depends on the family environment and early sexual experience of the child. What is inherited by an individual is the greater probability of acquiring homophilia under the conditions permitting its development.

If the kin-selection hypothesis is correct, homosexual behavior is likely still to be associated with role specialization and the favoring of kin in hunter-gatherer and simple agricultural societies, in other words those contemporary cultures most similar to the ones in which human social behavior evolved genetically during prehistory. The connection appears to exist. In some of the more primitive cultures that survived long enough to be studied by anthropologists, male homosexuals were berdaches, individuals who adopted women's dress and manner and who even married other men. They often became shamans, powerful members of the group able to influence its key decisions, or were specialized in some other way, in women's work, matchmaking, peacemaking, or as advisors to the tribal leaders. The female counterparts of berdaches are also known but are less well documented. It is further true that in western industrial societies, homosexual men score higher than heterosexuals on intelligence tests and are upwardly mobile to an exceptional degree. They select white collar professions disproportionately and regardless of their initial socioeconomic status are prone to enter specialties in which they deal directly with other people. They are more successful on the average within their chosen professions. Finally, apart from the difficulties created by the disapproval of their sexual preferences, homosexuals are considered by others to be generally well adapted in social relationships.

All of this information amounts to little more than a set of clues. It is not decisive by the usual canons of science. A great deal of additional, careful research is needed. But the clues are enough to establish that the traditional Judeo-Christian view of homosexual behavior is inadequate and probably wrong. The assumptions of this religion-sanctioned hypothesis have lain hidden for centuries but can now be exposed and tested by objective standards. I believe it entirely correct to say that the kin-selection hypothesis is more consistent with the existing evidence.

The juxtaposition of biology and ethics in the case of homosexuality requires sensitivity and care. It would be inappropriate to consider homosexuals as a separate genetic caste, however beneficent their historic and contemporary roles might prove to be. It would be even more illogical, and unfortunate, to make past genetic adaptedness a necessary criterion for current acceptance. But it would be tragic to continue to discriminate against homosexuals on the basis of religious dogma supported by the unlikely assumption that they are biologically unnatural.

The central argument of this chapter has been that human sexuality can be much more precisely defined with the aid of the new advances in evolutionary theory. To omit this mode of reasoning is to leave us blind to an important part of our history, the ultimate meaning of our behavior, and the significance of the choices that lie before us.

Through the instruments of education and law, each society must make a series of choices concerning sexual discrimination, the standards of sexual behavior, and the reinforcement of the family. As government and technology become more complex and interdependent, the choices have to be correspondingly precise and sophisticated. One way or the other, intuitively or with the aid of science, evolutionary history will be entered in the calculations, because human nature is stubborn and cannot be forced without a cost.

There is a cost, which no one can yet measure, awaiting the society that moves either from juridical equality of opportunity between the sexes to a statistical equality of their performance in the professions, or back toward deliberate sexual discrimination. Another unknown cost awaits the society that decides to reorganize itself into smoothly functioning nuclear families, or to abolish families in favor of communal kibbutzim. There is still another cost — and some of our members are already paying it in personal suffering — for the society that insists on conformity to a particular range of heterosexual practices. We believe that cultures can be rationally designed. We can teach and reward and coerce. But in so doing we must also consider the price of each culture, measured in the time and energy required for training and enforcement and in the less tangible currency of human happiness that must be spent to circumvent our innate predispositions.

Chapter 7. Altruism

"The blood of martyrs is the seed of the church." With that chilling dictum the third-century theologian Tertullian confessed the fundamental flaw of human altruism, an intimation that the purpose of sacrifice is to raise one human group over another. Generosity without hope of reciprocation is the rarest and most cherished of human behaviors, subtle and difficult to define, distributed in a highly selective pattern, surrounded by ritual and circumstance, and honored by medallions and emotional orations. We sanctify true altruism in order to reward it and thus to make it less than true, and by that means to promote its recurrence in others. Human altruism, in short, is riddled to its foundations with the expected mammalian ambivalence.

As mammals would be and ants would not, we are fascinated by the extreme forms of self-sacrifice. In the First and Second World Wars, Korea, and Vietnam, a large percentage of Congressional Medals of Honor were awarded to men who threw themselves on top of grenades to shield comrades, aided the rescue of others from battle sites at the cost of certain death to themselves, or made other extraordinary decisions that led to the same fatal end. Such altruistic suicide is the ultimate act of courage and emphatically deserves the country's highest honor. But it is still a great puzzle. What could possibly go on in the minds of these men in the moment of desperation? "Personal vanity and pride are always important factors in situations of this kind," James Jones wrote in *WWII*,

> and the sheer excitement of battle can often lead a man to death willingly, where without it he might have balked. But in the absolute, ultimate end, when your final extinction is right there only a few yards farther on staring back at you, there may be a sort of penultimate national, and social, and even racial, masochism — a sort of hotly joyous, almost-sexual enjoyment and acceptance — which keeps you going the last few steps. The ultimate luxury of just *not giving a damn* any more.

The annihilating mixture of reason and passion, which has been described often in first-hand accounts of the battlefield, is only the extreme phenomenon that lies beyond the innumerable smaller impulses of courage and generosity that bind societies together. One is tempted to leave the matter there, to accept the purest elements of altruism as simply the better side of human nature. Perhaps, to put the best possible construction on the matter, conscious altruism is a transcendental quality that distinguishes human beings from animals. But scientists are not accustomed to declaring any phenomenon off limits, and it is precisely through the deeper analysis of altruism that sociobiology seems best prepared at this time to make a novel contribution.

I doubt if any higher animal, such as an eagle or a lion, has ever deserved a Congressional Medal of Honor by the ennobling criteria used in our society. Yet minor altruism does occur frequently, in forms instantly understandable in human terms, and is bestowed not just on offspring but on other members of the species as well. Certain small birds, robins, thrushes and titmice, for example, warn others of the approach of a hawk. They crouch low and emit a distinctive thin, reedy whistle. Although the warning call has acoustic properties that make its source difficult to locate in space, to whistle at all seems at the very least unselfish; the caller would be wiser not to betray its presence but rather to remain silent.

Other than man, chimpanzees may be the most altruistic of all mammals. In addition to sharing meat after their cooperative hunts, they also practice adoption. Jane Goodall has observed three cases at the Gombe Stream National Park in Tanzania, all involving orphaned infants taken over by adult brothers and sisters. It is of considerable interest, for more theoretical reasons to be discussed shortly, that the altruistic behavior was displayed by the closest possible relatives rather than by experienced females with children of their own, females who might have supplied the orphans with milk and more adequate social protection.

In spite of a fair abundance of such examples among vertebrates, it is only in the lower animals, and in the social insects particularly, that we encounter altruistic suicide comparable to man's. Many members of ant, bee, and wasp colonies are ready to defend their nests with insane charges against intruders. This is the reason that people move with circumspection around honeybee hives and yellow-jacket burrows, but can afford to relax near the nests of solitary species such as sweat bees and mud daubers.

The social stingless bees of the tropics swarm over the heads of human beings who venture too close and lock their jaws so tightly onto tufts of hair that their bodies are pulled loose from their heads when they are combed out. Some species pour a burning glandular secretion onto the skin during these sacrificial attacks. In Brazil, they are called *cagafogos* ("fire defecators"). The great entomologist William Morton Wheeler described an encounter with the "terrible bees," during which they removed patches of skin from his face, as the worst experience of his life.

Honeybee workers have stings lined with reversed barbs like those on fishhooks. When a bee attacks an intruder at the hive, the sting catches in the skin; as the bee moves away, the sting remains embedded, pulling out the entire venom gland and much of the viscera with it. The bee soon dies, but its attack has been more effective than if it withdrew the sting intact. The reason is that the venom gland continues to leak poison into the wound, while a bananalike odor emanating from the base of the sting incites other members of the hive to launch kamikaze attacks of their own at the same spot. From the point of view of the colony as a whole, the suicide of an individual accomplishes more than it loses. The total worker force consists of twenty thousand to eighty thousand members, all sisters born from eggs laid by the mother queen. Each bee has a natural life span of only about fifty days, after which it dies of old age. So to give a life is only a little thing, with no genes being spilled.

My favorite example among the social insects is provided by an African termite with the orotund technical name *Globitermes sulfureus*. Members of this species' soldier caste are quite literally walking bombs. Huge paired glands extend from their heads back through most of their bodies. When they attack ants and other enemies, they eject a yellow glandular secretion through their mouths; it congeals in the air and often fatally entangles both the soldiers and their antagonists. The spray appears to be powered by contractions of the muscles in the abdominal wall. Sometimes the contractions become so violent that the abdomen and gland explode, spraying the defensive fluid in all directions.

Sharing the capacity for extreme sacrifice does not mean that the human mind and the "mind" of an insect (if such exists) work alike. But it does mean that the impulse need not be ruled divine or otherwise transcendental, and we are justified in seeking a more conventional biological explanation. A basic problem immediately arises in connection with such an explanation: fallen heroes do not have children. If self-sacrifice results in fewer descendants, the genes that allow heroes to be created can be expected to disappear gradually from the population. A narrow interpretation of Darwinian natural selection would predict this outcome: because people governed by selfish genes must prevail over those with altruistic genes, there should also be a tendency over many generations for selfish genes to increase in prevalence and for a population to become ever less capable of responding altruistically.

How then does altruism persist? In the case of social insects, there is no doubt at all. Natural selection has been broadened to include kin selection. The self-sacrificing termite soldier protects the rest of its colony, including the queen and king, its parents. As a result, the soldier's more fertile brothers and sisters flourish, and through them the altruistic genes are multiplied by a greater production of nephews and nieces.

It is natural, then, to ask whether through kin selection the capacity for altruism has also evolved in human beings. In other words, do the emotions we feel, which in exceptional individuals may climax in total self-sacrifice, stem ultimately from hereditary units that were implanted by the favoring of relatives during a period of hundreds or thousands of generations? This explanation gains some strength from the circumstance that during most of mankind's history the predominant social unit was the immediate family and a tight network of other close relatives. Such exceptional cohesion, combined with detailed kin classifications made possible by high intelligence, might explain why kin selection has been more forceful in human beings than in monkeys and other mammals.

To anticipate a common objection raised by many social scientists and others, let me grant at once that the form and intensity of altruistic acts are to a large extent culturally determined. Human social evolution is obviously more cultural than genetic. The point is that the underlying emotion, powerfully manifested in virtually all human societies, is what is considered to evolve through genes. The sociobiological hypothesis does not therefore account for differences among societies, but it can explain why human beings differ from other mammals and why, in one narrow aspect, they more closely resemble social insects.

The evolutionary theory of human altruism is greatly complicated by the ultimately self-serving quality of most forms of that altruism. No sustained form of human altruism is explicitly and totally self-annihilating. Lives of the most towering heroism are paid out in the expectation of great reward, not the least of which is a belief in personal immortality. When poets speak of happy acquiescence in death they do not mean death at all but apotheosis, or nirvana; they revert to what Yeats called the artifice of eternity. Near the end of *Pilgrim's Progress* we learn of the approaching death of Valiant-for-Truth:

> Then said he, "I am going to my fathers, and though with great difficulty I am got hither, yet now I do not repent me of all the trouble I have been at to arrive where I am. My sword, I give to him that shall succeed me in my pilgrimage, and my courage and skill, to him that can get it. My marks and my scars I carry with me, to be a witness for me that I have fought his battles who now will be my rewarder."

Valiant-for-Truth then utters his last words, *Grave where is thy victory?*, and departs as his friends hear trumpets sounded for him on the other side.

Compassion is selective and often ultimately self-serving. Hinduism permits lavish preoccupation with the self and close relatives but does not encourage compassion for unrelated individuals or, least of all, outcastes. A central goal of Nibbanic Buddhism is preserving the individual through altruism. The devotee earns points toward a better personal life by performing generous acts and offsets bad acts with meritorious ones. While embracing the concept of universal compassion, both Buddhist and Christian countries have found it expedient to wage aggressive wars, many of which they justify in the name of religion.

Compassion is flexible and eminently adaptable to political reality; that is to say it conforms to the best interests of self, family, and allies of the moment. The Palestinian refugees have received the sympathy of the world and have been the beneficiaries of rage among the Arab nations. But little is said about the Arabs killed by King Hussein or those who live in Arab countries with fewer civil rights and under far worse material conditions than the displaced people of the West Bank. When Bangladesh began its move toward independence in 1971, the President of Pakistan unleashed the Punjabi army in a campaign of terror that ultimately cost the lives of a million Bengalis and drove 9.8 million others into exile. In this war more Moslem people were killed or driven from their homes than make up the entire populations of Syria and Jordan. Yet not a single Arab state, conservative or radical, supported the Bangladesh struggle for independence. Most denounced the Bengalis while proclaiming Islamic solidarity with West Pakistan.

To understand this strange selectivity and resolve the puzzle of human altruism we must distinguish two basic forms of cooperative behavior. The altruistic impulse can be irrational and unilaterally directed at others; the bestower expresses no desire for equal return and performs no unconscious actions leading to the same end. I have called this form of behavior "hard-core" altruism, a set of responses relatively unaffected by social reward or punishment beyond childhood. Where such behavior exists, it is likely to have evolved through kin selection or natural selection operating on entire, competing family or tribal units. We would expect hard-core altruism to serve the altruist's closest relatives and to decline steeply in frequency and intensity as relationship becomes more distant. "Soft-core" altruism, in contrast, is ultimately selfish. The "altruist" expects reciprocation from society for himself or his closest relatives. His good behavior is calculating, often in a wholly conscious way, and his maneuvers are orchestrated by the excruciatingly intricate sanctions and demands of society. The capacity for soft-core altruism can be expected to have evolved primarily by selection of individuals and to be deeply influenced by the vagaries of cultural evolution. Its psychological vehicles are lying, pretense, and deceit, including self-deceit, because the actor is most convincing who believes that his performance is real.

A key question of social theory, then, must be the relative amounts of hard-core as opposed to soft-core altruism. In honeybees and termites, the issue has already been settled: kin selection is paramount, and altruism is virtually all hard-core. There are no hypocrites among the social insects. This tendency also prevails among the higher animals. It is true that a small amount of reciprocation is practiced by monkeys and apes. When male anubis baboons struggle for dominance, they sometimes solicit one another's aid. A male stands next to an enemy and a friend and swivels his gaze back and forth between the two while continuously threatening the enemy. Baboons allied in this manner are able to exclude solitary males during competition for estrous females. Despite the obvious advantages of such arrangements, however, coalitions are the rare exception in baboons and other intelligent animals.

But in human beings soft-core altruism has been carried to elaborate extremes. Reciprocation among distantly related or unrelated individuals is the key to human society. The perfection of the social contract has broken the ancient vertebrate constraints imposed by rigid kin selection. Through the convention of reciprocation, combined with a flexible, endlessly productive language and a genius for verbal classification, human beings fashion long-remembered agreements upon which cultures and civilizations can be built.

Yet the question remains: Is there a foundation of hard-core altruism beneath all of this contractual superstructure? The conception is reminiscent of David Hume's striking conjecture that reason is the slave of the passions. So we ask, to what biological end are the contracts made, and just how stubborn is nepotism?

The distinction is important because pure, hard-core altruism based on kin selection is the enemy of civilization. If human beings are to a large extent guided by programmed learning rules and canalized emotional development to favor their own relatives and tribe, only a limited amount of global harmony is possible. International cooperation will approach an upper limit, from which it will be knocked down by the perturbations of war and economic struggle, canceling each upward surge based on pure reason. The imperatives of blood and territory will be the passions to which reason is slave. One can imagine genius continuing to serve biological ends even after it has disclosed and fully explained the evolutionary roots of unreason.

My own estimate of the relative proportions of hard-core and soft-core altruism in human behavior is optimistic. Human beings appear to be sufficiently selfish and calculating to be capable of indefinitely greater harmony and social homeostasis. This statement is not self-contradictory. True selfishness, if obedient to the other constraints of mammalian biology, is the key to a more nearly perfect social contract.

My optimism is based on evidence concerning the nature of tribalism and ethnicity. If altruism were rigidly unilateral, kin and ethnic ties would be maintained with commensurate tenacity. The lines of allegiance, being difficult or impossible to break, would become progressively tangled until cultural change was halted in their snarl. Under such circumstances the preservation of social units of intermediate size, the extended family and the tribe, would be paramount. We should see it working at the conspicuous expense of individual welfare on the one side and of national interest on the other.

In order to understand this idea more clearly, return with me for a moment to the basic theory of evolution. Imagine a spectrum of self-serving behavior. At one extreme only the individual is meant to benefit, then the nuclear family, next the extended family (including cousins, grandparents, and others who might play a role in kin selection), then the band, the tribe, chiefdoms, and finally, at the other extreme, the highest sociopolitical units. Which units along this spectrum are most favored by the innate predispositions of human social behavior? To reach an answer we can look at natural selection from another perspective: those units subjected to the most intense natural selection, those that reproduce and die most frequently and in concert with the demands of the environment, will be the ones protected by the innate behavior of individual organisms belonging to them. In sharks natural selection occurs overwhelmingly at the individual level; all behavior is self-centered and exquisitely appropriate to the welfare of one shark and its immediate offspring. In the Portuguese man-of-war and other siphonophore jellyfish that consist of great masses of highly coordinated individuals, the unit of selection is almost exclusively the colony. The individual organism, a zooid reduced and compacted into the gelatinous mass, counts for very little. Some members of the colony lack stomachs, others lack nervous systems, most never reproduce, and almost all can be shed and regenerated. Honeybees, termites, and other social insects are only slightly less colony-centered.

Human beings obviously occupy a position on the spectrum somewhere between the two extremes, but exactly where? The evidence suggests to me that human beings are well over toward the individual end of the spectrum. We are not in the position of sharks, or selfish monkeys and apes, but we are closer to them than we are to honeybees in this single parameter. Individual behavior, including seemingly altruistic acts bestowed on tribe and nation, are directed, sometimes very circuitously, toward the Darwinian advantage of the solitary human being and his closest relatives. The most elaborate forms of social organization, despite their outward appearance, serve ultimately as the vehicles of individual welfare. Human altruism appears to be substantially hard-core when directed at closest relatives, although still to a much lesser degree than in the case of the social insects and the colonial invertebrates. The remainder of our altruism is essentially soft. The predicted result is a melange of ambivalence, deceit, and guilt that continuously troubles the individual mind.

The same intuitive conclusion has been drawn independently by the biologist Robert L. Trivers and in less technical terms by the social psychologist Donald T. Campbell, who has been responsible for a renaissance of interest in the scientific study of human altruism and moral behavior. And in reviewing a large body of additional information from sociology, Milton M. Gordon has generalized that "man defending the honor or welfare of his ethnic group is man defending himself."

The primacy of egocentrism over race has been most clearly revealed by the behavior of ethnic groups placed under varying conditions of stress. For example, Sephardic Jews from Jamaica who emigrate to England or America may, according to personal circumstances, remain fully Jewish by joining the Jews of the host society, or may abandon their ethnic ties promptly, marry gentiles, and blend into the host culture. Puerto Ricans who migrate back and forth between San Juan and New York are even more versatile. A black Puerto Rican behaves as a member of the black minority in Puerto Rico and as a member of the Puerto Rican minority in New York. If given the opportunity to use affirmative action in New York he may emphasize his blackness. But in personal relationships with whites he is likely to minimize the color of his skin by references to his Spanish language and Latin culture. And like Sephardic Jews, many of the better educated Puerto Ricans sever their ethnic ties and quickly penetrate the mainland culture.

Orlando Patterson of Harvard University has shown how such behavior in the melting pot, when properly analyzed, can lead to general insights concerning human nature itself. The Caribbean Chinese are an example of an ethnic group whose history resembles a controlled experiment. By examining their experience closely we may distinguish some of the key cultural variables affecting ethnic allegiance. When the Chinese immigrants arrived in Jamaica in the late nineteenth century they were presented with the opportunity to occupy and dominate the retail system. An economic vacuum existed: the black peasantry was still tied to a rural existence centered on the old slave plantations, while the white Jews and gentiles constituted an upper class who regarded retailing as beneath them. The hybrid "coloreds" might have filled the niche but did not, because they were anxious to imitate the whites into whose socioeconomic class they hoped to move. The Chinese were a tiny minority of less than one percent, yet they were able to take over retail trade in Jamaica and to improve their lot enormously. They did it by simultaneously specializing in trade and consolidating their ranks through ethnic allegiance and restrictive marriage customs. Racial consciousness and deliberate cultural exclusiveness were put to the service of individual welfare.

In the 1950s the social environment changed drastically, and with it the Chinese ethos. When Jamaica became independent, the new ruling elite were a racial mixture firmly committed to a national, synthetic Creole culture. It now was in the best interests of the Chinese enclave to join the elite socially, and they did so with alacrity. Within fifteen years they ceased to be a distinct cultural group. They altered their mode of business from mostly wholesaling to the construction and management of supermarkets and shopping plazas. They adopted the bourgeois life style and Creole culture and shifted emphasis from the traditional extended family to the nuclear family. Through it all they maintained racial consciousness, not as a blind genetic imperative but as an economic strategy. The most success-ful families had always been the most endogamous ones; women were the means by which wealth was exchanged, consolidated, and kept within small family groups. Because the custom did not inter-fere with assimilation into the rest of Creole culture, the Jamaican Chinese kept it.

In Guyana, the small country on the northern coast of South America formerly known as British Guiana, the Chinese immigrants faced a very different kind of challenge, although their background was the same as that of their Jamaican counterparts. They had been brought to the colony from the same parts of China as the Jamaican Chinese and to a large extent by the same agent. But in the towns of old British Guiana they found the retail trade already filled by an-other ethnic group, the Portuguese, who had arrived during the 1840s and 1850s. The white ruling class favored the Portuguese as the group racially and culturally closer to themselves. Some Chinese did enter the retail trade, but they were never overwhelmingly suc-cessful. Others were forced to enter other occupations, including governmental positions. None of these alternatives conferred the same advantage on ethnic awareness; it was not possible, as in the retail trade, to maximize earnings through ethnic exclusiveness. And so the Chinese of British Guiana eagerly joined the emerging Creole culture. By 1915 one of their keenest observers, Cecil Clementi, could say, "British Guiana possesses a Chinese society of which China knows nothing, and to which China is almost unknown." But their success was more than compensatory: although the Chinese make up only 0.6 percent of the total population, they are now pow-erful elements of the middle class, and from their ranks came the first president of the republic, Arthur Chung.

From his own Caribbean research, and from comparable studies by other sociologists, Patterson has drawn three conclusions about allegiance and altruism: (1) When historical circumstances bring the interests of race, class, and ethnic membership into conflict, the individual maneuvers to achieve the least amount of conflict. (2) As a rule the individual maneuvers so as to optimize his own interests over all others. (3) Although racial and ethnic interests may prevail temporarily, socioeconomic classes are paramount in the long run.

The strength and scope of an individual's ethnic identity are determined by the general interests of his socioeconomic class, and they serve the interests of, first, himself, then his class, and finally his ethnic group. There is a convergent principle in political science known as Director's Law, which states that income in a society is distributed to the benefit of the class that controls the government. In the United States this is of course the middle class. And it can be further noted that all kinds of institutions, from corporations to churches, evolve in a way that promotes the best interests of those who control them. Human altruism, to come back to the biological frame of reference, is soft. To search for hard elements, one must probe very close to the individual, and no further away than his children and a few other closest kin.

Yet it is a remarkable fact that all human altruism is shaped by powerful emotional controls of the kind intuitively expected to occur in its hardest forms. Moral aggression is most intensely expressed in the enforcement of reciprocation. The cheat, the turncoat, the apostate, and the traitor are objects of universal hatred. Honor and loyalty are reinforced by the stiffest codes. It seems probable that learning rules, based on innate, primary reinforcement, lead human beings to acquire these values and not others with reference to members of their own group. The rules are the symmetrical counterparts to the canalized development of territoriality and xenophobia, which are the equally emotional attitudes directed toward members of other groups.

I will go further to speculate that the deep structure of altruistic behavior, based on learning rules and emotional safeguards, is rigid and universal. It generates a set of predictable group responses of the kind that have been catalogued in more technical works such as those prepared by Bernard Berelson, Robert A. LeVine, Nathan Glazer, and other social scientists. One such generalization is the following: the poorer the ingroup, the more it uses group narcissism as a form of compensation. Another: the larger the group, the weaker the narcissistic gratification that individuals obtain by identifying with it, the less cohesive the group bonds, and the more likely individuals are to identify with smaller groups inside the group. And still another: if subgroups of some kind already exist, a region that appears homogeneous while still part of a larger country is not likely to remain so if it becomes independent. Most inhabitants of such regions respond to narrowing of political boundaries by narrowing the focus of their group identification.

In summary, soft-core altruism is characterized by strong emotion and protean allegiance. Human beings are consistent in their codes of honor but endlessly fickle with reference to whom the codes apply. The genius of human sociality is in fact the ease with which alliances are formed, broken, and reconstituted, always with strong emotional appeals to rules believed to be absolute. The important distinction is today, as it appears to have been since the Ice Age, between the ingroup and the outgroup, but the precise location of the dividing line is shifted back and forth with ease. Professional sports thrive on the durability of this basic phenomenon. For an hour or so the spectator can resolve his world into an elemental physical struggle between tribal surrogates. The athletes come from everywhere and are sold and traded on an almost yearly basis. The teams themselves are sold from city to city. But it does not matter; the fan identifies with an aggressive ingroup, admires teamwork, bravery, and sacrifice, and shares the exultation of victory.

Nations play by the same rules. During the past thirty years geopolitical alignments have changed from a confrontation between the

Axis and the Allies to one between the Communists and the Free World, then to oppositions between largely economic blocs. The United Nations is both a forum for the most idealistic rhetoric of humankind and a kaleidoscope of quickly shifting alliances based on selfish interests.

The mind is simultaneously puzzled by the cross-cutting struggles of religion. Some Arab extremists think the struggle against Israel is a jihad for the sacred cause of Islam. Christian evangelists forge an alliance with God and his angels against the hosts of Satan to prepare the world for the Second Coming. It was instructive to see Eldridge Cleaver, the one-time revolutionary, and Charles Colson, the archetypal secret agent, lift themselves out of their old epistemic frameworks and move to the side of Christ on this more ancient battleground of religion. The substance matters little, the form is all.

It is exquisitely human to make spiritual commitments that are absolute to the very moment they are broken. People invest great energies in arranging their alliances while keeping other, equally cathectic options available. So long as the altruistic impulse is so powerful, it is fortunate that it is also mostly soft. If it were hard, history might be one great hymenopterous intrigue of nepotism and racism, and the future bleak beyond endurance. Human beings would be eager, literally and horribly, to sacrifice themselves for their blood kin. Instead, there is in us a flawed capacity for a social contract, mammalian in its limitations, combined with a perpetually renewing, optimistic cynicism with which rational people can accomplish a great deal.

We return then to the property of hypertrophy, the cultural inflation of innate human properties. Malcolm Muggeridge once asked me, What about Mother Theresa? How can biology account for the living saints among us? Mother Theresa, a member of the

Missionaries of Charity, cares for the desperately poor of Calcutta; she gathers the dying from the sidewalks, rescues abandoned babies from garbage dumps, attends the wounds and diseases of people no one else will touch. Despite international recognition and rich awards, Mother Theresa lives a life of total poverty and grinding hard work. In *Something Beautiful for God*, Muggeridge wrote of his feelings after observing her closely in Calcutta: "Each day Mother Theresa meets Jesus; first at the Mass, whence she derives sustenance and strength; then in each needing, suffering soul she sees and tends. They are one and the same Jesus; at the altar and in the streets. Neither exists without the other."

Can culture alter human behavior to approach altruistic perfection? Might it be possible to touch some magical talisman or design a Skinnerian technology that creates a race of saints? The answer is no. In sobering reflection, let us recall the words of Mark's Jesus: "Go forth to every part of the world, and proclaim the Good News to the whole creation. Those who believe it and receive baptism will find salvation; those who do not believe will be condemned." There lies the fountainhead of religious altruism. Virtually identical formulations, equally pure in tone and perfect with respect to ingroup altruism, have been urged by the seers of every major religion, not omitting Marxism-Leninism. All have contended for supremacy over others. Mother Theresa is an extraordinary person but it should not be forgotten that she is secure in the service of Christ and the knowledge of her Church's immortality. Lenin, who preached a no less utopian, if rival, covenant, called Christianity unutterably vile and a contagion of the most abominable kind; that compliment has been returned many times by Christian theologians.

"If only it were all so simple!," Aleksandr Solzhenitsyn wrote in *The Gulag Archipelago.* "If only there were evil people somewhere insidiously committing evil deeds, and it were necessary only to sep-

arate them from the rest of us and destroy them. But the line dividing good and evil cuts through the heart of every human being. And
who is willing to destroy a piece of his own heart?"

Sainthood is not so much the hypertrophy of human altruism as its
ossification. It is cheerfully subordinate to the biological imperatives
above which it is supposed to rise. The true humanization of altruism, in the sense of adding wisdom and insight to the social contract,
can come only through a deeper scientific examination of morality.
Lawrence Kohlberg, an educational psychologist, has traced what
he believes to be six sequential stages of ethical reasoning through
which each person progresses as part of his normal mental development. The child moves from an unquestioning dependence on external rules and controls to an increasingly sophisticated set of in-
ternalized standards, as follows: (1) simple obedience to rules and
authority to avoid punishment, (2) conformity to group behavior
to obtain rewards and exchange favors, (3) good-boy orientation,
conformity to avoid dislike and rejection by others, (4) duty orientation, conformity to avoid censure by authority, disruption of
order, and resulting guilt, (5) legalistic orientation, recognition of
the value of contracts, some arbitrariness in rule formation to maintain the common good, (6) conscience or principle orientation, primary allegiance to principles of choice, which can overrule law in
cases the law is judged to do more harm than good.

The stages were based on children's verbal responses, as elicited
by questions about moral problems. Depending on intelligence and
training, individuals can stop at any rung on the ladder. Most attain
stages four or five. By stage four they are at approximately the level
of morality reached by baboon and chimpanzee troops. At stage five,
when the ethical reference becomes partly contractual and legalistic,
they incorporate the morality on which I believe most of human social evolution has been based. To the extent that this interpretation
is correct, the ontogeny of moral development is likely to have been

genetically assimilated and is now part of the automatically guided process of mental development. Individuals are steered by learning rules and relatively inflexible emotional responses to progress through stage five. Some are diverted by extraordinary events at critical junctures. Sociopaths do exist. But the great majority of people reach stages four or five and are thus prepared to exist harmoniously — in Pleistocene hunter-gatherer camps.

Since we no longer live as small bands of hunter-gatherers, stage six is the most nearly nonbiological and hence susceptible to the greatest amount of hypertrophy. The individual selects principles against which the group and the law are judged. Precepts chosen by intuition based on emotion are primarily biological in origin and are likely to do no more than reinforce the primitive social arrangements. Such a morality is unconsciously shaped to give new rationalizations for the consecration of the group, the proselytizing role of altruism, and the defense of territory.

But to the extent that principles are chosen by knowledge and reason remote from biology, they can at least in theory be non-Darwinian. This leads us ineluctably back to the second great spiritual dilemma. The philosophical question of interest that it generates is the following: Can the cultural evolution of higher ethical values gain a direction and momentum of its own and completely replace genetic evolution? I think not. The genes hold culture on a leash. The leash is very long, but inevitably values will be constrained in accordance with their effects on the human gene pool. The brain is a product of evolution. Human behavior — like the deepest capacities for emotional response which drive and guide it — is the circuitous technique by which human genetic material has been and will be kept intact. Morality has no other demonstrable ultimate function.

Chapter 8. Religion

The predisposition to religious belief is the most complex and powerful force in the human mind and in all probability an ineradicable part of human nature. Emile Durkheim, an agnostic, characterized religious practice as the consecration of the group and the core of society. It is one of the universals of social behavior, taking recognizable form in every society from hunter-gatherer bands to socialist republics. Its rudiments go back at least to the bone altars and funerary rites of Neanderthal man. At Shanidar, Iraq, sixty thousand years ago, Neanderthal people decorated a grave with seven species of flowers having medicinal and economic value, perhaps to honor a shaman. Since that time, according to the anthropologist Anthony F. C. Wallace, mankind has produced on the order of 100 thousand religions.

Skeptics continue to nourish the belief that science and learning will banish religion, which they consider to be no more than a tissue of illusions. The noblest among them are sure that humanity migrates toward knowledge by logotaxis, an automatic orientation toward information, so that organized religion must continue its retreat as darkness before enlightenment's brightening dawn. But this conception of human nature, with roots going back to Aristotle and

Zeno, has never seemed so futile as today. If anything, knowledge is being enthusiastically harnessed to the service of religion. The United States, technologically and scientifically the most sophisticated nation in history, is also the second most religious — after India. According to a Gallup poll taken in 1977, 94 percent of Americans believe in God or some form of higher being, while 31 percent have undergone a moment of sudden religious insight or awakening, their brush with the epiphany. The most successful book in 1975 was Billy Graham's *Angels: God's Secret Messengers*, which sold 810 thousand hard-cover copies.

In the Soviet Union, organized religion still flourishes and may even be undergoing a small renaissance after sixty years of official discouragement. In a total population of 250 million, at least thirty million are members of the Orthodox Church — twice the number in the Communist Party — five million are Roman Catholics and Lutherans, and another two million belong to evangelical sects such as the Baptists, Pentecostals, and Seventh-Day Adventists. Still another twenty to thirty million are Moslems, while 2.5 million belong to that most resilient of all groups, Orthodox Jews. Thus, institutionalized Soviet Marxism, which is itself a form of religion embellished with handsome trappings, has failed to displace what many Russians for centuries have considered the soul of their national existence.

Scientific humanism has done no better. In his *System of Positive Polity*, published between 1846 and 1854, Auguste Comte argued that religious superstition can be defeated at its source. He recommended that educated people fabricate a secular religion consisting of hierarchies, liturgy, canons, and sacraments not unlike those of Roman Catholicism, but with society replacing God as the Grand Being to worship. Today, scientists and other scholars, organized into learned groups such as the American Humanist Society and Institute on Religion in an Age of Science, support little magazines distributed by subscription and organize campaigns to discredit Chris-

tian fundamentalism, astrology, and Immanuel Velikovsky. Their
crisply logical salvos, endorsed by whole arrogances of Nobel
Laureates, pass like steel-jacketed bullets through fog. The humanists are vastly outnumbered by true believers, by the people who
follow Jeane Dixon but have never heard of Ralph Wendell Burhoe. Men, it appears, would rather believe than know. They would
rather have the void as purpose, as Nietzsche despairingly wrote so
long ago when science was at its full promise, than be void of purpose.

Other well-meaning scholars have tried to reconcile science and
religion by compartmentalizing the two rivals. Newton saw himself
not only as a scientist but as a historical scholar whose duty was to
decipher the Scriptures as a true historical record. Although his own
mighty effort created the first modern synthesis of the physical sciences, he regarded that achievement as only a way station to an understanding of the supernatural. The Creator, he believed, has given
the scholar two works to read, the book of nature and the book of
scriptures. Today, thanks to the relentless advance of the science
which Newton pioneered, God's immanence has been pushed to
somewhere below the subatomic particles or beyond the farthest visible galaxy. This apparent exclusion has spurred still other philosophers and scientists to create "process theology," in which God's
presence is inferred from the inherent properties of atomic structure.
As conceived originally by Alfred North Whitehead, God is not to
be viewed as an extraneous force, who creates miracles and presides
over the metaphysical verities. He is present continuously and ubiq-
uitously. He covertly guides the emergence of molecules from
atoms, living organisms from molecules, and mind from matter. The
properties of the electron cannot be finally announced until their
end product, the mind, is understood. Process is reality, reality process, and the hand of God is manifest in the laws of science. Hence
religious and scientific pursuits are intrinsically compatible, so that

well-meaning scientists can return to their calling in a state of mental peace. But all this, the reader will immediately recognize, is a world apart from the real religion of the aboriginal corroboree and the Council of Trent.

Today, as always before, the mind cannot comprehend the meaning of the collision between irresistible scientific materialism and immovable religious faith. We try to cope through a step-by-step pragmatism. Our schizophrenic societies progress by knowledge but survive on inspiration derived from the very beliefs which that knowledge erodes. I suggest that the paradox can be at least intellectually resolved, not all at once but eventually and with consequences difficult to predict, if we pay due attention to the sociobiology of religion. Although the manifestations of the religious experience are resplendent and multidimensional, and so complicated that the finest of psychoanalysts and philosophers get lost in their labyrinth, I believe that religious practices can be mapped onto the two dimensions of genetic advantage and evolutionary change.

Let me moderate this statement at once by conceding that if the principles of evolutionary theory do indeed contain theology's Rosetta stone, the translation cannot be expected to encompass in detail all religious phenomena. By traditional methods of reduction and analysis science can explain religion but cannot diminish the importance of its substance.

A historical episode will serve as a parable in the sociobiology of religion. The aboriginal people of Tasmania, like the exotic marsupial wolves that once shared their forest habitat, are extinct. It took the British colonists only forty years to finish them off (the wolves lasted another hundred years, to 1950). This abruptness is especially unfortunate from the viewpoint of anthropology, because the Tasmanians — the "wild ones" — had no chance to transmit even a description of their culture to the rest of the world. Little is known be-

yond the fact that they were hunters and gatherers of small stature
with reddish-brown skin and frizzled hair, and, according to the explorers who first encountered them, an open and happy temperament. Their origin can only be guessed. Most probably they were
the descendents of aboriginal Australians who reached Tasmania
about ten thousand years ago, then adapted biologically and culturally to the cool, wet forests of the island. We are left with only a few
photographs and skeletons. Not even the language can be reconstructed, because few Europeans who met the Tasmanians thought
it worthwhile to take notes.

The British settlers who began arriving in the early 1800s regarded the Tasmanians as something less than human. They were
only little brown obstacles to agriculture and civilization. Accord-
ingly, they were rounded up during organized hunts and murdered
for slight offenses. One party of men, women, and children was cut
down by gunfire simply for running in the direction of whites during one of the kangaroo hunts conducted en masse by the aborigines.
Most died of syphilis and other European diseases. The point of no
return was reached by 1842, when the number of Tasmanians had
dwindled from an original five thousand or so to fewer than thirty.
The women were then too old to have any more children, and the
culture had atrophied.

The last stages of the aboriginals' decline was presided over by a
remarkable altruist, George Robinson, a missionary from London.
In 1830, when several hundred Tasmanians remained, Robinson began a heroic and virtually single-handed attempt to save the race.
By approaching the hunted survivors sympathetically, he persuaded
them to follow him out of their forest retreats into surrender. A few
then settled in the new towns of the settlers, where they invariably
became derelicts. The rest were taken by Robinson to a reserve on
Flinders Island, an isolated post northeast of Tasmania. There they

were fed salt beef and sweet tea, dressed in European clothes, and instructed in personal hygiene, money changing, and strict Calvinism. The old culture was then completely forbidden to them.

Each day the Tasmanians went to their little church to hear a sermon by George Robinson. From this terminal phase of their cultural history we do have records, rendered in pidgin English: "One God . . . Native good, native dead, go to sky . . . Bad native dead, goes down, evil spirit, fire stops. Native cry, cry, cry . . ." The catechism repeated the easily comprehended message:

> What will God do to the world by and by?
>   Burn it!
> Do you like the Devil?
>   No!
> What did God make us for?
>   His own purposes . . .

The Tasmanians could not survive the harsh smelting of their souls. They grew somber and lethargic and ceased producing children. Many died from influenza and pneumonia. Finally the remnants were moved to a new reserve near Hobart, on the mainland of Tasmania. The last male, known as King Billy to the Europeans, died in 1869, and the several remaining old women followed a few years later. They were the objects of intense curiosity and, finally, respect. During this period George Robinson raised a large family of his own. His life's goal had been to try to retrieve the Tasmanians from extinction, by substituting in good conscience the more civilized form of religious subjugation for murder. Yet by the stark biological algorithm that guided him unconsciously, Robinson was not a failure.

While growing increasingly sophisticated, anthropology and history continue to support Max Weber's conclusion that the more elementary religions seek the supernatural for purely mundane rewards: long life, abundant land and food, averting physical catastrophes, and the conquest of enemies. A kind of cultural Darwinism also operates during the competition among sects in the evolution of more advanced religions. Those that gain adherents grow; those that cannot, disappear. Consequently religions are like other human institutions in that they evolve in directions that enhance the welfare of the practitioners. Because this demographic benefit must accrue to the group as a whole, it can be gained partly by altruism and partly by exploitation, with certain sectors profiting at the expense of others. Alternatively, the benefit can arise as the sum of the generally increased fitnesses of all of the members. The resulting distinction in social terms is between the more oppressive and the more beneficent religions. All religions are probably oppressive to some degree, especially when they are promoted by chiefdoms and states. There is a principle in ecology, Gause's law, which states that maximum competition is to be found between those species with identical needs. In a similar manner, the one form of altruism that religions seldom display is tolerance of other religions. Their hostility intensifies when societies clash, because religion is superbly serviceable to the purposes of warfare and economic exploitation. The conqueror's religion becomes a sword, that of the conquered a shield.

Religion constitutes the greatest challenge to human sociobiology and its most exciting opportunity to progress as a truly original theoretical discipline. If the mind is to any extent guided by Kantian imperatives, they are more likely to be found in religious feeling than in rational thought. Even if there is a materialist basis of religious process and it lies within the grasp of conventional science, it will be difficult to decipher for two reasons.

First, religion is one of the major categories of behavior undeniably unique to the human species. The principles of behavioral evolution drawn from existing population biology and experimental studies on lower animals are unlikely to apply in any direct fashion to religion.

Second, the key learning rules and their ultimate, genetic motivation are probably hidden from the conscious mind, because religion is above all the process by which individuals are persuaded to subordinate their immediate self-interest to the interests of the group. Votaries are expected to make short-term physiological sacrifices for their own long-term genetic gains. Self-deception by shamans and priests perfects their own performance and enhances the deception practiced on their constituents. In the midst of absurdity the trumpet is certain. Decisions are automatic and quick, there being no rational calculus by which groups of individuals can compute their inclusive genetic fitness on a day-to-day basis and thus *know* the amount of conformity and zeal that is optimum for each act. Human beings require simple rules that solve complex problems, and they tend to resist any attempt to dissect the unconscious order and resolve of their daily lives. The principle has been expressed in psychoanalytic theory by Ernest Jones as follows: "Whenever an individual considers a given (mental) process as being too obvious to permit of any investigation into its origin, and shows resistance to such an investigation, we are right in suspecting that the actual origin is concealed from him — almost certainly on account of its unacceptable nature."

The deep structure of religious belief can be probed by examining natural selection at three successive levels. At the surface, selection is ecclesiastic: rituals and conventions are chosen by religious leaders for their emotional impact under contemporary social conditions. Ecclesiastic selection can be either dogmatic and stabilizing or evangelistic and dynamic. In either case the results are culturally transmitted; hence variations in religious practice from one society to the next are based on learning and not on genes. At the next level selection is ecological. Whatever the fidelity of ecclesiastic selection to the emotions of the faithful, however easily its favored conventions are learned, the resulting practice must eventually be tested by the demands of the environment. If religions weaken their societies during warfare, encourage the destruction of the environment, shorten lives, or interfere with procreation they will, regardless of their short-term emotional benefits, initiate their own decline. Finally, in the midst of these complicated epicycles of cultural evolution and population fluctuation, the frequencies of genes are changing.

The hypothesis before us is that some gene frequencies are changed in consistent ways by ecclesiastic selection. Human genes, it will be recalled, program the functioning of the nervous, sensory, and hormonal systems of the body, and thereby almost certainly influence the learning process. They constrain the maturation of some behaviors and the learning rules of other behaviors. Incest taboos, taboos in general, xenophobia, the dichotomization of objects into the sacred and profane, nosism, hierarchical dominance systems, intense attention toward leaders, charisma, trophyism, and trance-induction are among the elements of religious behavior most likely to be shaped by developmental programs and learning rules. All of these processes act to circumscribe a social group and bind its members together in unquestioning allegiance. Our hypothesis requires that such constraints exist, that they have a physiological basis, and that the physiological basis in turn has a genetic origin. It implies that ecclesiastical choices are influenced by the chain of events that lead from the genes through physiology to constrained learning during single lifetimes.

According to the hypothesis, the frequencies of the genes themselves are reciprocally altered by the descending sequence of several kinds of selection — ecclesiastic, ecological, and genetic — over many lifetimes. Religious practices that consistently enhance survival and procreation of the practitioners will propagate the physiological controls that favor acquisition of the practices during single lifetimes. The genes that prescribe the controls will also be favored.

Because religious practices are remote from the genes during the development of individual human beings, they may vary widely during cultural evolution. It is even possible for groups, such as the Shakers, to adopt conventions that reduce genetic fitness for as long as one or a few generations. But over many generations, the underlying genes will pay for their permissiveness by declining in the population as a whole. Other genes governing mechanisms that resist decline of fitness produced by cultural evolution will prevail, and the deviant practices will disappear. Thus culture relentlessly tests the controlling genes, but the most it can do is to replace one set of genes with another.

This hypothesis of interaction between genes and culture can be either supported or disproved if we examine the effects of religion at the ecological and genetic levels. By far the more accessible is the ecological. We need to ask: What are the effects of each religious practice on the welfare of individuals and tribes? How did the practice originate in history and under what environmental circumstances? To the extent that it represents a response to necessity or has improved the efficiency of a society over many generations, the correlation conforms to the interaction hypothesis. To the extent that it runs counter to these expectations, even if it cannot be related to reproductive fitness in a relatively simple, reasonable way, the hypothesis is in difficulty. Finally, the genetically programmed constraints on learning revealed by developmental psychology must prove to be consistent with the major trends in religious practice. If they are not, the hypothesis is doubtful, and it can be legitimately supposed that in this case cultural evolution has mimicked the theoretically predicted pattern of genetic evolution.

In order to pursue the investigation over a sufficiently wide array of topics, the definition of religious behavior must be broadened to include magic and the more sanctified tribal rituals, as well as the more elaborate beliefs constructed around mythology. I believe that even when this step is taken, the evidence is consistent with the hypothesis of gene-culture interaction, and few episodes in the history of religion contravene it.

Consider ritual. Stirred by an early enthusiasm for Lorenz-Tinbergen ethology, some social scientists drew an analogy between human ceremonies and the displays of animal communication. The comparison is at best imprecise. Most animal displays are discrete signals that convey limited meaning. They are commensurate with the postures, facial expressions, and elementary sounds of human nonlinguistic communication. A few animal displays, such as the most complex forms of sexual advertisement and bond formation in birds, are so impressively elaborate that they have occasionally been termed ceremonies by zoologists. But even here the comparison is misleading. Most human rituals have more than just an immediate signal value. As Durkheim stressed, they not only label but reaffirm and rejuvenate the moral values of the community.

The sacred rituals are the most distinctively human. Their elementary forms are concerned with magic, the active attempt to manipulate nature and the gods. Upper Paleolithic art from the caves of Western Europe indicates a preoccupation with game animals. There are many scenes showing spears and arrows embedded in the bodies of the prey. Other drawings depict men dancing in animal disguises or standing with heads bowed in front of animals. Probably the function was sympathetic magic, derived from the notion that what is done with an image will come to pass with the real thing. The anticipatory action is comparable to the intention movements of animals, which in the course of evolution have often been ritualized into communicative signals. The waggle dance of the honeybee is actually a miniaturized rehearsal of the flight from the hive to the food. The "straight run" performed as the middle piece of the figure-eight dance is varied precisely in direction and duration to convey the magnitude of these parameters in the true flight to follow. Primitive man would have understood the meaning of such complex animal behavior easily. Magic was, and still is in some societies, practiced by special people variously called shamans, sorcerers, or medicine men. They alone were believed to have the secret knowledge and power to deal with the supernatural forces of nature, and as such their influence sometimes exceeded that of the tribal headmen.

As the anthropologist Roy A. Rappaport has shown in a recent critical review of the subject, sacred rites mobilize and display primitive societies in ways that appear to be directly and biologically advantageous. Ceremonies can offer information on the strength and wealth of tribes and families. Among the Maring of New Guinea, there are no chiefs or other leaders who command allegiance during war. A group gives a ritual dance, and individual men indicate their willingness to lend military support by whether they attend the dance or not. The strength of the consortium is then precisely determined by a head count. In more advanced societies military parades, embellished by the paraphernalia and rituals of the state religion, serve the same purpose. The famous potlatch ceremonies of the Northwest Coast Indians enable individuals to advertise their wealth by the amount of goods they give away. Leaders are further able to mobilize the energies of groups of kin into the manufacture of surplus goods, enlarging the power of families.

Rituals also regularize relationships in which there would otherwise be ambiguity and wasteful imprecision. The best examples of this mode of communication are rites of passage. As a boy matures his transition from child to man is very gradual in a biological and psychological sense. There will be times when he behaves as a child whereas an adult response would have been more appropriate, and vice versa. Society has difficulty in classifying him one way or the other. The rite of passage eliminates this ambiguity by arbitrarily changing the classification from a continuous gradient into a dichotomy. It also serves to cement the ties of the young person to the adult group that accepts him.

The proneness of the human mind to attack problems by binary classification is also manifested in witchcraft. The psychological etiology of witchcraft has been reconstructed with great skill by social scientists such as Robert A. LeVine, Keith Thomas, and Monica Wilson. The immediate motivations revealed by their studies are partly emotional and partly rational. In all societies the shaman is in a position either to heal or to cast malevolent spells. So long as his role is unchallenged, he and his kin enjoy added power. If his actions are not only benevolent but also sanctioned through ritual, they contribute to the resolve and integration of the society. The biological advantages of institutionalized witchcraft therefore seem clear.

The witchhunt, the obverse of sorcery's practice, is a much more puzzling phenomenon and provides a truly interesting challenge to our theoretical investigation. Why do people from time to time declare themselves bewitched, or their society afflicted, and search for malevolent supernatural powers in their neighbors? Exorcisms and inquisitions are phenomena as complex and powerful as the practice of magic, but even here motivations prove to be rooted in the self-seeking of individuals. The epidemic of witchhunting in Tudor and Stuart England is one of the better documented examples. Before this period (1560–1680) the Catholic Church had offered the citizenry a well-organized system of ritual precautions against evil spirits and malevolent spells. The Church had, in effect, practiced positive witchcraft. The Reformation removed this psychological protection. Protestant ministers denounced the old religious practices while reaffirming the existence of evil magic. Deprived of ritual countermeasures, bewitched persons turned to the suspected witches themselves, accused them publicly, and sought their destruction.

A close examination of the court records has revealed the probable deeper motivation behind the persecutions. Typically the accuser had turned down a poor woman who asked for food or some other favor and was then struck by a personal misfortune such as a crop failure or death in the family. By fastening the blame on the woman the accuser accomplished two purposes. He took direct action against what he sincerely believed to be the cause of his troubles, in obedience to a certain logic that recognized the apartness and meddlesome behavior of alleged witches. The second motivation is more subtle and less easily proved. According to Thomas,

> The conflict between resentment and a sense of obligation produced the ambivalence which made it possible for men to turn begging women brusquely from the door and yet to suffer torment of conscience after having done so. The ensuing guilt was fertile ground for witchcraft accusations, since subsequent misfortune could be seen as retaliation on the part of the witch. The tensions that produced witchcraft allegations were those generated by a society which no longer held a clear view as to how its dependent members should be treated; they reflected the ethical conflict between the twin and opposing doctrines that those who did not work should not eat, and that it was blessed for the rich to support the poor.

So by transmuting the dilemma into a war against evil spirits, the accuser rationalized the more selfish course of action.

Among the Nyansongan of Kenya, witches are identified through gossip rather than by formal denunciation. The Nyansongan leaders, including the homestead heads, elders, chiefs, and members of the tribunal courts, usually reject the stories of witchcraft and attempt to resolve the disputes by discussion and arbitration. The looseness of the procedure permits individuals to peddle rumors and accusations as a means of calling attention to their personal problems.

The practical nature of witchcraft and other forms of magic is the reason such activities are often distinguished from the higher strata of "true" religion. Most scholars have followed Durkheim in making a fundamental distinction between the sacred, the core of religion, and the profane, the quality invested in magic and ordinary life. To sanctify a procedure or a statement is to certify it as beyond question and imply punishment for anyone who dares to contradict it. In the Hindu creation myths, for example, those who marry outside their caste go to the hellish Yama's kingdom after death, where they are forced to embrace red-hot human forms. So removed is the sacred from the profane that simply to speak of it in the wrong circumstances is a transgression. The sacred rites engender awe, an intimation of qualities beyond human understanding.

This extreme form of certification is granted to the practices and dogmas that serve the vital interests of the group. The individual is prepared by the sacred rituals for supreme effort and self-sacrifice. Overwhelmed by shibboleths, special costumes, and sacred dancing and music accurately keyed to his emotive centers, he is transformed by a religious experience. The votary is ready to reassert allegiance to his tribe and family, perform charities, consecrate his life, leave for the hunt, join the battle, die for God and country. It was true in the past, as John Pfeiffer has said:

> Everything they knew and believed, the full force of ancestral authority and tradition, came to a growing white-heat focus in ceremony. What began with a shaman performing in a trance among people around camp fires culminated in spectacles conducted by high priests and their cohorts from platforms elevated above the multitude. There was singing and chanting, words said over and over again, recited in singsong metrical patterns with punctuating rhymes at the ends of lines. Music, setting the pace in the background and echoing and rising to crescendos and climaxes, reinforced the beat. Dancers with

masks kept time to the words and the music as they acted out the roles of gods and heroes. Spectators moved with the rhythms and chanted ritual responses.

And so it continues to the present time, in usually more fragmented and muted versions. The modern traditionalist heresy of Catholicism and the evangelistic and revitalization movements of the Protestants are efforts to reverse the corroding secularization of society and to return to the old forms. An unthinking submission to the communal will remains among the most emotionally potent virtues among "good" people in the mainstream of the society. "Jesus is the answer" is the contemporary equivalent of *Deus vult*, the rallying cry of the First Crusade. God wills it, whatever the action, however hard the path. Mao Tse-tung said "We must persevere and work unceasingly, and we, too, will touch God's heart. Our God is none other than the Chinese people." When the gods are served, the Darwinian fitness of the members of the tribe is the ultimate if unrecognized beneficiary. We must now inquire: Is the readiness to be indoctrinated a neurologically based learning rule that evolved through the selection of clans competing one against the other?

In support of this simple biological hypothesis is the fact that the blinding force of religious allegiance can operate in the absence of theology. The May Day rallies of T'ien An Men Square would have been instantly understood by the Mayan multitudes, Lenin's tomb by the worshipers of Christ's bloodied shroud. Consider the following reflection by Grigori Pyatakov, one of Lenin's closest disciples: "A real Communist, that is, a man raised in the Party and who has absorbed its spirit becomes himself in a way a miracle man. For such a Party a true Bolshevik will readily cast out from his mind ideas in which he has believed for years. A true Bolshevik has submerged his personality in the collectivity, the 'Party,' to such an extent that he can make the necessary effort to break away from his own opinions and convictions, and can honestly agree with the Party — that is the test of a true Bolshevik."

In *The Denial of Death*, Ernest Becker reminds us that the guru phenomenon is a device for surrendering the self to a powerful and benevolent force. The Zen master demands absolute allegiance in every technique — the exact headstand, the exact manner of breathing — until the apprentice is drawn from the self and sustained by a magical power. The Zen archer no longer shoots the arrow; the interior of nature breaks into the world through the archer's perfect selflessness and releases the string.

The self-fulfilling cults of the present day, including Esalen, est, Arica, and scientology, are the vulgar replacements of the traditional forms. Their leaders receive a degree of obedience from otherwise intelligent Americans that would wring smiles of admiration from the most fanatical Sufi *shaykh*. In the Erhard Training Seminars (est), novitiates are pounded from the lectern with simplistic truths from the behavioral sciences and Eastern philosophy while being simultaneously bullied and soothed by attendants. They are not allowed to leave their seats to eat or go to the bathroom or even to stand and stretch. The reward, according to Peter Marin's personal study, is the masochistic relief that results from placing oneself into the hands of a master to whom omnipotence has been granted.

Advantage can accrue to both the individual and the society from such willing subordination. It was Henri Bergson who first recognized what might be the ultimate agent behind the mechanisms of emotional gratification. The extreme plasticity of human social behavior, Bergson noted, is both a great strength and a danger. If each family worked out its own rules of behavior, the society as a whole would disintegrate into chaos. To counteract selfish behavior and the dissolving power of high intelligence and idiosyncracy, each society must codify itself. Within broad limits any set of conventions works better than none at all. Because arbitrary codes work, organizations tend to be inefficient and marred by unnecessary inequities. As Rappaport has succinctly expressed it: "Sanctification transforms the arbitrary into the necessary, and regulatory mechanisms which are arbitrary are likely to be sanctified."

But the arbitrariness of sanctification engenders criticism, and within the more liberal and self-conscious societies visionaries and revolutionaries set out to change the system. Their ultimate purpose is to elevate codes of their own devising. Reform meets repression, because to the extent that the reigning code has been sanctified and mythologized, the majority of the people regard it as beyond question, and disagreement is defined as blasphemy.

The stage is thus set for the conflict of natural selection at the individual and group levels. In addressing this conflict we have come full circle to the theoretical question of the origin of altruism. Grant for the moment that there is a genetic predisposition to conformity and consecration. Was it installed by selection at the level of entire societies or by selection at the level of the individual? The question can be rephrased at the level of psychology: Is the behavior hard-core, programmed to safeguard the interests of the entire community, or is it soft-core and thereby prone to manipulation in the self-interest of individuals?

At one extreme, the one more likely to produce hard religiosity, the group is the unit of selection. When conformity becomes too weak, groups suffer decline and perhaps even extinction. In this hypothetical version it is still possible for selfish, individualistic members to gain the upper hand and to multiply at the expense of others. But the rising influence of their deviant predispositions accelerates the vulnerability of the society and hastens its decline. Societies with higher frequencies of such individuals, and hence of the genes that predispose to them, will give way to those less weakened in "genetic resolve," and the overall frequency of conforming individuals in the population as a whole will rise. The genetic capacity for blind conformity spreads at the expense of the genetic incapacity. Even the potential for self-sacrifice can be strengthened in this manner, because the willingness of individuals to relinquish rewards or even surrender their own lives will favor group survival. The loss of genes suffered through the deaths of disciplined individuals can be more than balanced by a gain of genes attained through expansion of the benefited group.

At the other extreme, generating a softer and more ambivalent religiosity, individual selection is the ruling force in Darwinian evolution. The ability of individuals to conform permits them to enjoy the benefits of membership with a minimum of energy expenditure and risk, and their behavior is sustained over long periods of time as the social norm. Although the rivals of the conformists in the society may gain a momentary advantage through selfishness and irreverence, it is lost in the long run through ostracism and repression. The conformists perform altruistic acts possibly to the extent of risking their own lives not because of a genetic predisposition selected through competition among entire societies, but because the group is occasionally able to take advantage of the indoctrinability which on other occasions is favorable to the individual.

These two possibilities need not be mutually exclusive; group and individual selection can be reinforcing. If success of the group requires spartan virtues and self-denying religiosity, victory can more than recompense the surviving faithful in land, power, and the opportunity to reproduce. The average individual will win this Darwinian game, and his gamble will be profitable, because the summed efforts of the participants give the average member a more than compensatory edge:

> The LORD spoke to Moses and said, "Count all that has been captured, man or beast, you and Eleazar the priest and the heads

of families in the community, and divide it equally between the fighting men who went on the campaign and the whole community. You shall levy a tax for the LORD: from the combatants it shall be one out of every five hundred, whether men, cattle, asses, or sheep, to be taken out of their share and given to Eleazar the priest as a contribution for the LORD. Out of the share of the Israelites it shall be one out of every fifty taken, whether man or beast, cattle, asses, or sheep, to be given to the Levites who are in charge of the LORD's tabernacle. (Num. 30:25–38)

The highest forms of religious practice, when examined more closely, can be seen to confer biological advantage. Above all they congeal identity. In the midst of the chaotic and potentially disorienting experiences each person undergoes daily, religion classifies him, provides him with unquestioned membership in a group claiming great powers, and by this means gives him a driving purpose in life compatible with his self-interest. His strength is the strength of the group, his guide the sacred covenant. The theologian and sociologist Hans J. Mol has aptly termed this key process the "sacralization of identity." The mind is predisposed — one can speculate that learning rules are physiologically programmed — to participate in a few processes of sacralization which in combination generate the institutions of organized religion.

The first mechanism is objectification, the description of reality with images and definitions that are easily understood and invulnerable to contradictions and exceptions. Heaven and hell, human life as an arena for the struggle between the forces of good and evil, gods controlling each force of nature, and spirits ready to enforce the taboos are examples of this device. Objectification creates an attractive framework on which to festoon symbols and myths.

Commitment is the second process of religion-making. The faithful consecrate their lives to the ideas that have been objectified and to the welfare of those who do the same. Commitment is pure tribalism enacted through emotional self-surrender. Its focus is on the mystic covenant and the shamans and priests whose translation of the codes is deemed necessary for certification. Commitment is attained by ceremonies, in which the arbitrary rules and sacred objects are consecrated and repetitively defined until they seem as much a part of human nature as love or hunger.

Finally there is myth: the narratives by which the tribe's special place in the world is explained in rational terms consistent with the listener's understanding of the physical world. Preliterate hunter-gatherers tell believable sacred stories about the creation of the world. Human beings and animals with supernatural powers and a special relationship to the tribe fight, eat, and beget offspring. Their actions explain a little bit of how nature works and why the tribe has a favored position on earth. The complexity of the myths increases with that of societies. They duplicate the essential structure in more fantastic forms. Tribes of demigods and heroes, warring for kingship and possession of territory, allocate dominion over different parts of the lives of mortal men. Over and again the myths strike the Manichaean theme of two supernal forces struggling for control of the world of man. For some of the Amerinds of the Amazon-Orinoco forests, for example, the contenders are two brothers representing the sun and the moon, one a benevolent creator, the other a trickster. In the later Hindu myths Brahma, benevolent lord of the universe, creates Night. She gives birth to the rakshasas, who try to eat Brahma and to destroy mortal men. Another recurrent theme in the more elaborate mythologies is the apocalypse and millenium, wherein it is forecast that the struggles will cease when a god descends to end the existing world and to create a new order.

Belief in such high gods is not universal. Among eighty-one hunter-gatherer societies surveyed by John W. M. Whiting, only twenty-eight, or 35 percent, included high gods in their sacred traditions. The concept of an active, moral God who created the world is even less widespread. Furthermore, this concept most commonly arises with a pastoral way of life. The greater the dependence on herding, the more likely the belief in a shepherd god of the Judeo-Christian type. In other kinds of society the belief occurs in 10 percent or less of those whose religion is known.

The God of monotheistic religions is always male; this strong patriarchal tendency has several cultural sources. Pastoral societies are highly mobile, tightly organized, and often militant, all features that tip the balance toward male authority. It is also significant that herding, the main economic base, is primarily the responsibility of men. Because the Hebrews were originally a herding people, the Bible describes God as a shepherd and the chosen people as his sheep. Islam, one of the strictest of all monotheistic faiths, grew to early power among the herding people of the Arabian peninsula.

The sociobiological explanation of faith in God leads to the crux of the role of mythology in modern life. It is obvious that human beings are still largely ruled by myth. Furthermore, much of contemporary intellectual and political strife is due to the conflict between three great mythologies: Marxism, traditional religion, and scientific materialism. Marxism is still regarded by purists as a form of scientific materialism, but it is not. The perception of history as an inevitable class struggle proceeding to the emergence of a lightly governed egalitarian society with production in control of the workers is supposed to be based on an understanding of the subterranean forces of pure economic process. In fact, it is equally based on an inaccurate interpretation of human nature. Marx, Engels, and all the disciples and deviationists after them, however sophisticated, have operated on a set of larger hidden premises about the deeper desires of human beings and the extent to which human behavior can be molded by social environments. These premises have never been tested. To the extent that they can be made explicit, they are inadequate or simply wrong. They have become the hidden wards of the historicist dogma they were supposed to generate.

Marxism is sociobiology without biology. The strongest opposition to the scientific study of human nature has come from a small number of Marxist biologists and anthropologists who are committed to the view that human behavior arises from a very few unstructured drives. They believe that nothing exists in the untrained human mind that cannot be readily channeled to the purposes of the revolutionary socialist state. When faced with the evidence of greater structure, their response has been to declare human nature off limits to further scientific investigation. A few otherwise very able scholars have gone so far as to suggest that merely to talk about the subject is dangerous, at least to their concept of progress. I hope that I have been able to show that this perception is profoundly wrong. At the same time, anxiety about the health of Marxism as a theory and a belief system *is* justified. Although Marxism was formulated as the enemy of ignorance and superstition, to the extent that it has become dogmatic it has faltered in that commitment and is now mortally threatened by the discoveries of human sociobiology.

But if Marxism is only an inaccurate product of scientific materialism, a failed satrap so to speak, traditional religion is not. As science proceeds to dismantle the ancient mythic stories one by one, theology retreats to the final redoubt from which it can never be driven. This is the idea of God in the creation myth: God as will, the cause of existence, and the agent who generated all of the energy in the original fireball and set the natural laws by which the universe evolved. So long as the redoubt exists, theology can slip out through its portals and make occasional sallies back into the real world.

Whenever other philosophers let their guard down, deists can, in the manner of process theology, postulate a pervasive transcendental will. They can even hypothesize miracles.

But make no mistake about the power of scientific materialism. It presents the human mind with an alternative mythology that until now has always, point for point in zones of conflict, defeated traditional religion. Its narrative form is the epic: the evolution of the universe from the big bang of fifteen billion years ago through the origin of the elements and celestial bodies to the beginnings of life on earth. The evolutionary epic is mythology in the sense that the laws it adduces here and now are believed but can never be definitely proved to form a cause-and-effect continuum from physics to the social sciences, from this world to all other worlds in the visible universe, and backward through time to the beginning of the universe. Every part of existence is considered to be obedient to physical laws requiring no external control. The scientist's devotion to parsimony in explanation excludes the divine spirit and other extraneous agents. Most importantly, we have come to the crucial stage in the history of biology when religion itself is subject to the explanations of the natural sciences. As I have tried to show, sociobiology can account for the very origin of mythology by the principle of natural selection acting on the genetically evolving material structure of the human brain.

If this interpretation is correct, the final decisive edge enjoyed by scientific naturalism will come from its capacity to explain traditional religion, its chief competitor, as a wholly material phenomenon. Theology is not likely to survive as an independent intellectual discipline. But religion itself will endure for a long time as a vital force in society. Like the mythical giant Antaeus who drew energy from his mother, the earth, religion cannot be defeated by those who merely cast it down. The spiritual weakness of scientific naturalism is due to the fact that it has no such primal source of power. While explaining the biological sources of religious emotional strength, it is unable in its present form to draw on them, because the evolutionary epic denies immortality to the individual and divine privilege to the society, and it suggests only an existential meaning for the human species. Humanists will never enjoy the hot pleasures of spiritual conversion and self-surrender; scientists cannot in all honesty serve as priests. So the time has come to ask: Does a way exist to divert the power of religion into the services of the great new enterprise that lays bare the sources of that power? We have come back at last to the second dilemma in a form that demands an answer.

Chapter 9. Hope

The first dilemma has been created by the seemingly fatal deterioration of the myths of traditional religion and its secular equivalents, principal among which are ideologies based on a Marxian interpretation of history. The price of these failures has been a loss of moral consensus, a greater sense of helplessness about the human condition and a shrinking of concern back toward the self and the immediate future. The intellectual solution of the first dilemma can be achieved by a deeper and more courageous examination of human nature that combines the findings of biology with those of the social sciences. The mind will be more precisely explained as an epiphenomenon of the neuronal machinery of the brain. That machinery is in turn the product of genetic evolution by natural selection acting on human populations for hundreds of thousands of years in their ancient environments. By a judicious extension of the methods and ideas of neurobiology, ethology, and sociobiology a proper foundation can be laid for the social sciences, and the discontinuity still separating the natural sciences on the one side and the social sciences and humanities on the other might be erased.

If this solution to the first dilemma proves even partially correct, it will lead directly to the second dilemma: the conscious choices that must be made among our innate mental propensities. The elements of human nature are the learning rules, emotional reinforcers, and hormonal feedback loops that guide the development of social behavior into certain channels as opposed to others. Human nature is not just the array of outcomes attained in existing societies. It is also the potential array that might be achieved through conscious design by future societies. By looking over the realized social systems of hundreds of animal species and deriving the principles by which these systems have evolved, we can be certain that all human choices represent only a tiny subset of those theoretically possible. Human nature is, moreover, a hodgepodge of special genetic adaptations to an environment largely vanished, the world of the Ice-Age hunter-gatherer. Modern life, as rich and rapidly changing as it appears to those caught in it, is nevertheless only a mosaic of cultural hypertrophies of the archaic behavioral adaptations. And at the center of the second dilemma is found a circularity: we are forced to choose among the elements of human nature by reference to value systems which these same elements created in an evolutionary age now long vanished.

Fortunately, this circularity of the human predicament is not so tight that it cannot be broken through an exercise of will. The principal task of human biology is to identify and to measure the constraints that influence the decisions of ethical philosophers and everyone else, and to infer their significance through neurophysiological and phylogenetic reconstructions of the mind. This enterprise is a necessary complement to the continued study of cultural evolution. It will alter the foundation of the social sciences but in no way diminish their richness and importance. In the process it will fashion a biology of ethics, which will make possible the selection of a more deeply understood and enduring code of moral values.

In the beginning the new ethicists will want to ponder the cardinal value of the survival of human genes in the form of a common pool over generations. Few persons realize the true consequences of the dissolving action of sexual reproduction and the corresponding un-importance of "lines" of descent. The DNA of an individual is made up of about equal contributions of all the ancestors in any given generation, and it will be divided about equally among all descendants at any future moment. All of us have more than two hundred ancestors who were living in 1700—each of whom contributed far less than one chromosome to the living descendant—and, depending on the amount of outbreeding that took place, up to millions in 1066. Henry Adams put it nicely for those of Norman-English descent when he noted that if "we could go back and live again in all our two hundred and fifty million arithmetical ancestors of the eleventh century, we should find ourselves doing many surprising things, but among the rest we should certainly be ploughing most of the fields of the Contentin and Calvados; going to mass in every parish church in Normandy; rendering military service to every lord, spiritual or temporal, in all this region; and helping to build the Abbey Church at Mont-Saint-Michel." Go back another few thousands of years—only a tick in the evolutionary clock—and the gene pool from which one modern Briton has emerged spreads over Europe, to North Africa, the Middle East, and beyond. The individual is an evanescent combination of genes drawn from this pool, one whose hereditary material will soon be dissolved back into it. Because natural selection has acted on the behavior of individuals who benefit themselves and their immediate relatives, human nature bends us to the imperatives of selfishness and tribalism. But a more detached view of the long-range course of evolution should allow us to see beyond the blind decision-making process of natural selection and to envision the history and future of our own genes against the background of the entire human species. A word already in use intuitively defines this view: nobility. Had dinosaurs grasped the concept they might have survived. They might have been us.

I believe that a correct application of evolutionary theory also favors diversity in the gene pool as a cardinal value. If variation in mental and athletic ability is influenced to a moderate degree by heredity, as the evidence suggests, we should expect individuals of truly extraordinary capacity to emerge unexpectedly in otherwise undistinguished families, and then fail to transmit these qualities to their children. The biologist George C. Williams has written of such productions in plants and animals as Sisyphean genotypes; his reasoning is based on the following argument from elementary genetics. Almost all capacities are prescribed by combinations of genes at many sites on the chromosomes. Truly exceptional individuals, weak or strong, are, by definition, to be found at the extremes of statistical curves, and the hereditary substrate of their traits come together in rare combinations that arise from random processes in the formation of new sex cells and the fusion of sex cells to create new organisms. Since each individual produced by the sexual process contains a unique set of genes, very exceptional combinations of genes are unlikely to appear twice even within the same family. So if genius is to any extent hereditary, it winks on and off through the gene pool in a way that would be difficult to measure or predict. Like Sisyphus rolling his boulder up and over to the top of the hill only to have it tumble down again, the human gene pool creates hereditary genius in many ways in many places only to have it come apart the next generation. The genes of the Sisyphean combinations are probably spread throughout populations. For this reason alone, we are justified in considering the preservation of the entire gene pool as a contingent primary value until such time as an almost unimaginably greater knowledge of human heredity provides us with the option of a democratically contrived eugenics.

Universal human rights might properly be regarded as a third primary value. The idea is not general; it is largely the invention of recent European-American civilization. I suggest that we will want to give it primary status not because it is a divine ordinance (kings used to rule by divine right) or through obedience to an abstract principle of unknown extraneous origin, but because we are mammals. Our societies are based on the mammalian plan: the individual strives for personal reproductive success foremost and that of his immediate kin secondarily; further grudging cooperation represents a compromise struck in order to enjoy the benefits of group membership. A rational ant—let us imagine for a moment that ants and other social insects had succeeded in evolving high intelligence—would find such an arrangement biologically unsound and the very concept of individual freedom intrinsically evil. We will accede to universal rights because power is too fluid in advanced technological societies to circumvent this mammalian imperative; the long-term consequences of inequity will always be visibly dangerous to its temporary beneficiaries. I suggest that this is the true reason for the universal rights movement and that an understanding of its raw biological causation will be more compelling in the end than any rationalization contrived by culture to reinforce and euphemize it.

The search for values will then go beyond the utilitarian calculus of genetic fitness. Although natural selection has been the prime mover, it works through a cascade of decisions based on secondary values that have historically served as the enabling mechanisms for survival and reproductive success. These values are defined to a large extent by our most intense emotions: enthusiasm and a sharpening of the senses from exploration; exaltation from discovery; triumph in battle and competitive sports; the restful satisfaction from an altruistic act well and truly placed; the stirring of ethnic and national pride; the strength from family ties; and the secure biophilic pleasure from the nearness of animals and growing plants.

There is a neurophysiology of such responses to be deciphered, and their evolutionary history awaits reconstruction. A kind of principle of the conservation of energy operates among them, such that the emphasis of any one over others still retains the potential summed power of all. Poets have noted it well, as in the calm phrasing of Mary Barnard's Sappho:

> Some say a cavalry corps,
> some infantry, some, again,
> will maintain that the swift oars
>
> of our fleet are the finest
> sight on dark earth; but I say
> that whatever one loves, is.

Although the means to measure these energies are lacking, I suspect psychologists would agree that they can be rechanneled substantially without losing strength, that the mind fights to retain a certain level of order and emotional reward. Recent evidence suggests that dreams are produced when giant fibers in the brainstem fire upward through the brain during sleep, stirring the cerebral cortex to activity. In the absence of ordinary sensory information from the outside, the cortex responds by calling up images from the memory banks and fabricating plausible stories. In an analogous manner the mind will always create morality, religion, and mythology and empower them with emotional force. When blind ideologies and religious beliefs are stripped away, others are quickly manufactured as replacements. If the cerebral cortex is rigidly trained in the techniques of critical analysis and packed with tested information, it will reorder all that into some form of morality, religion, and mythology. If the mind is instructed that its pararational activity cannot be combined with the rational, it will divide itself into two compartments so that both activities can continue to flourish side by side.

This mythopoeic drive can be harnessed to learning and the rational search for human progress if we finally concede that scientific materialism is itself a mythology defined in the noble sense. So let me give again the reasons why I consider the scientific ethos superior to religion: its repeated triumphs in explaining and controlling the physical world; its self-correcting nature open to all competent to devise and conduct the tests; its readiness to examine all subjects sacred and profane; and now the possibility of explaining traditional religion by the mechanistic models of evolutionary biology. The last achievement will be crucial. If religion, including the dogmatic secular ideologies, can be systematically analyzed and explained as a product of the brain's evolution, its power as an external source of morality will be gone forever and the solution of the second dilemma will have become a practical necessity.

The core of scientific materialism is the evolutionary epic. Let me repeat its minimum claims: that the laws of the physical sciences are consistent with those of the biological and social sciences and can be linked in chains of causal explanation; that life and mind have a physical basis; that the world as we know it has evolved from earlier worlds obedient to the same laws; and that the visible universe today is everywhere subject to these materialist explanations. The epic can be indefinitely strengthened up and down the line, but its most sweeping assertions cannot be proved with finality.

What I am suggesting, in the end, is that the evolutionary epic is probably the best myth we will ever have. It can be adjusted until it comes as close to truth as the human mind is constructed to judge the truth. And if that is the case, the mythopoeic requirements of the mind must somehow be met by scientific materialism so as to reinvest our superb energies. There are ways of managing such a shift honestly and without dogma. One is to cultivate more intensely the relationship between the sciences and humanities. The great British biologist J. B. S. Haldane said of science and literature, "I am absolutely convinced that science is vastly more stimulating to the imagination than are the classics, but the products of the stimulus do not normally see the light because scientific men as a class are devoid of any perception of literary form." Indeed, the origin of the universe in the big bang of fifteen billion years ago, as deduced by astronomers and physicists, is far more awesome than the first chapter of Genesis or the Ninevite epic of Gilgamesh. When the scientists project physical processes backward to that moment with the aid of mathematical models they are talking about everything — literally everything — and when they move forward in time to pulsars, supernovas, and the collision of black holes they probe distances and mysteries beyond the imaginings of earlier generations. Recall how God lashed Job with concepts meant to overwhelm the human mind:

> Who is this whose ignorant words
> cloud my design in darkness?
> Brace yourself and stand up like a man;
> I will ask questions, and you shall answer . . .
> Have you descended to the springs of the sea
> or walked in the unfathomable deep?
> Have the gates of death been revealed to you?
> Have you ever seen the door-keepers of the place of darkness?
> Have you comprehended the vast expanse of the world?
> Come, tell me all this, if you know.

And yes, we *do* know and we have told. Jehovah's challenges have been met and scientists have pressed on to uncover and to solve even greater puzzles. The physical basis of life is known; we understand approximately how and when it started on earth. New species have been created in the laboratory and evolution has been traced at the molecular level. Genes can be spliced from one kind of organism into another. Molecular biologists have most of the knowledge needed to create elementary forms of life. Our machines, settled on Mars, have transmitted panoramic views and the results of chemical soil analysis. Could the Old Testament writers have conceived of such activity? And still the process of great scientific discovery gathers momentum.

Yet, astonishingly, the high culture of Western civilization exists largely apart from the natural sciences. In the United States intellectuals are virtually defined as those who work in the prevailing mode of the social sciences and humanities. Their reflections are devoid of the idioms of chemistry and biology, as though humankind were still in some sense a numinous spectator of physical reality. In the pages of *The New York Review of Books, Commentary, The New Republic, Daedalus, National Review, Saturday Review,* and other literary journals articles dominate that read as if most of basic science had halted during the nineteenth century. Their content consists largely of historical anecdotes, diachronic collating of outdated, verbalized theories of human behavior, and judgments of current events according to personal ideology — all enlivened by the pleasant but frustrating techniques of effervescence. Modern science is still regarded as a problem-solving activity and a set of technical marvels, the importance of which is to be valuated in an ethos extraneous to science. It is true that many "humanistic" scientists step outside scientific materialism to participate in the culture, sometimes as expert witnesses and sometimes as aspiring authors, but they almost never close the gap between the two worlds of discourse. With rare exceptions they are the tame scientists, the token emissaries of what must be viewed by their hosts as a barbaric culture still ungraced by a written language. They are degraded by the label they accept too readily: popularizers. Very few of the great writers, the ones who can trouble and move the deeper reaches of the mind, ever address real science on its own terms. Do they know the nature of the challenge?

The desired shift in attention could come more easily now that the human mind is subject to the network of causal explanation. Every epic needs a hero: the mind will do. Even astronomers, accustomed to thinking about ten billion galaxies and distances just short of infinity, must agree that the human brain is the most complex device that we know and the crossroads of investigation by every major natural science. The social scientists and humanistic scholars, not omitting theologians, will eventually have to concede that scientific naturalism is destined to alter the foundations of their systematic inquiry by redefining the mental process itself.

I began this book with an exposition of the often dialectic nature of scientific advance. The discipline abuts the antidiscipline; the antidiscipline succeeds in reordering the phenomena of the discipline by reduction to its more fundamental laws; but the new synthesis created in the discipline profoundly alters the antidiscipline as the interaction widens. I suggested that biology, and especially neurobiology and sociobiology, will serve as the antidiscipline of the social sciences. I will now go further and suggest that the scientific materialism embodied in biology will, through a reexamination of the mind and the foundations of social behavior, serve as a kind of antidiscipline to the humanities. No Comtian revolution will take place, no sudden creation of a primitively scientific culture. The translation will be gradual. In order to address the central issues of the humanities, including ideology and religious belief, science itself must become more sophisticated and in part specially crafted to deal with the peculiar features of human biology.

I hope that as this syncretism proceeds, a true sense of wonder will reinvade the broader culture. We need to speak more explicitly of the things we do not know. The epic of which natural scientists write in technical fragments still has immense gaps and absorbing mysteries, not the least of which is the physical basis of the mind. Like blank spaces on the map of a partly explored world, their near borders can be fixed but their inner magnitude only roughly guessed. Scientists and humanistic scholars can do far better than they have at articulating the great goals toward which literate people move as on a voyage of discovery. Unknown and surprising things await. They are as accessible as in those days of primitive wonder when the early European explorers went forth and came upon new worlds and the first microscopists watched bacteria swim across drops of water. As knowledge grows science must increasingly become the stimulus to imagination.

Such a view will undoubtedly be opposed as elitist by some who regard economic and social problems as everywhere overriding. There is an element of truth in that objection. Can anything really matter while people starve in the Sahel and India and rot in the prisons of Argentina and the Soviet Union? In response it can be asked, do we want to know, in depth and for all time, why we care? And when these problems are solved, what then? The stated purpose of governments everywhere is human fulfillment in some sense higher than animal survival. In almost all socialist revolutions the goals of highest priority, next to consecration of the revolution, are education, science, and technology — the combination that leads inexorably back to the first and second dilemmas.

This view will be rejected even more firmly by those whose emotional needs are satisfied by traditional organized religion. God and the church, they will claim, cannot be extinguished ex parte by a rival mythology based on science. They will be right. God remains a viable hypothesis as the prime mover, however undefinable and untestable that conception may be. The rituals of religion, especially the rites of passage and the sanctification of nationhood, are deeply entrenched and incorporate some of the most magnificent elements of existing cultures. They will certainly continue to be practiced long after their etiology has been disclosed. The anguish of death alone will be enough to keep them alive. It would be arrogant to suggest that a belief in a personal, moral God will disappear, just as it would be reckless to predict the forms that ritual will take as scientific materialism appropriates the mythopoeic energies to its own ends.

I also do not envision scientific generalization as a substitute for art or as anything more than a nourishing symbiont of art. The artist, including the creative writer, communicates his most personal experience and vision in a direct manner chosen to commit his audience emotionally to that perception. Science can hope to explain artists, and artistic genius, and even art, and it will increasingly use art to investigate human behavior, but it is not designed to transmit experience on a personal level or to reconstitute the full richness of the experience from the laws and principles which are its first concern by definition.

Above all, I am not suggesting that scientific naturalism be used as an alternative form of organized formal religion. My own reasoning follows in a direct line from the humanism of the Huxleys, Waddington, Monod, Pauli, Dobzhansky, Cattell, and others who have risked looking this Gorgon in the face. Each has achieved less than his purpose, I believe, for one or the other of two reasons. He has either rejected religious belief as animism or else recommended that it be sequestered in some gentle preserve of the mind where it can live out its culture-spawned existence apart from the mainstream of intellectual endeavor. Humanists show a touching faith in the power of knowledge and the idea of evolutionary progress over the minds of men. I am suggesting a modification of scientific humanism through the recognition that the mental processes of religious belief — consecration of personal and group identity, attention to charismatic leaders, mythopoeism, and others — represent programmed predispositions whose self-sufficient components were incorporated into the neural apparatus of the brain by thousands of generations of genetic evolution. As such they are powerful, ineradicable, and at the center of human social existence. They are also structured to a degree not previously appreciated by most philosophers. I suggest further that scientific materialism must accommodate them on two levels: as a scientific puzzle of great complexity and interest, and as a source of energies that can be shifted in new directions when scientific materialism itself is accepted as the more powerful mythology.

That transition will proceed at an accelerating rate. Man's destiny is to know, if only because societies with knowledge culturally dominate societies that lack it. Luddites and anti-intellectuals do not master the differential equations of thermodynamics or the biochemical cures of illness. They stay in thatched huts and die young. Cultures with unifying goals will learn more rapidly than those that lack them, and an autocatalytic growth of learning will follow because scientific materialism is the only mythology that can manufacture great goals from the sustained pursuit of pure knowledge.

I believe that a remarkable effect will be the increasingly precise specification of history. One of the great dreams of social theorists — Vico, Marx, Spencer, Spengler, Teggart, and Toynbee, among the most innovative — has been to devise laws of history that can foretell something of the future of mankind. Their schemes came to little because their understanding of human nature had no scientific basis; it was, to use a favored expression of scientific reporting, orders of magnitude too imprecise. The invisible hand remained invisible; the summed actions of thousands or millions of poorly understood individual human beings was not to be computed. Now there is reason to entertain the view that the culture of each society travels along one or the other of a set of evolutionary trajectories whose full array is constrained by the genetic rules of human nature. While broadly scattered from an anthropocentric point of view, this array still represents only a tiny subset of all the trajectories that would be possible in the absence of the genetic constraints.

As our knowledge of human nature grows, and we start to elect a system of values on a more objective basis, and our minds at last align with our hearts, the set of trajectories will narrow still more. We already know, to take two extreme and opposite examples, that the worlds of William Graham Sumner, the absolute Social Darwinist, and Mikhail Bakunin, the anarchist, are biologically impossible. As the social sciences mature into predictive disciplines, the permissible trajectories will not only diminish in number but our descendants will be able to sight farther along them.

Then mankind will face the third and perhaps final spiritual dilemma. Human genetics is now growing quickly along with all other branches of science. In time, much knowledge concerning the genetic foundation of social behavior will accumulate, and techniques may become available for altering gene complexes by molecular engineering and rapid selection through cloning. At the very least, slow evolutionary change will be feasible through conventional eugenics. The human species can change its own nature. What will it choose? Will it remain the same, teetering on a jerrybuilt foundation of partly obsolete Ice-Age adaptations? Or will it press on toward still higher intelligence and creativity, accompanied by a greater — or lesser — capacity for emotional response? New patterns of sociality could be installed in bits and pieces. It might be possible to imitate genetically the more nearly perfect nuclear family of the white-handed gibbon or the harmonious sisterhoods of the honeybees. But we are talking here about the very essence of humanity. Perhaps there is something already present in our nature that will prevent us from ever making such changes. In any case, and fortunately, this third dilemma belongs to later generations.

In the spirit of the enrichment of the evolutionary epic, modern writers often summon the classical mythic heroes to illustrate their view of the predicament of humankind: the existential Sisyphus, turning fate into the only means of expression open to him; hesitant

Arjuna at war with his conscience on the Field of Righteousness; disastrous Pandora bestowing the ills of mortal existence on human beings; and uncomplaining Atlas, steward of the finite Earth. Prometheus has gone somewhat out of fashion in recent years as a concession to resource limitation and managerial prudence. But we should not lose faith in him. Come back with me for a moment to the original, Aeschylean Prometheus:

> *Chorus*: Did you perhaps go further than you have told us?
> *Prometheus*: I caused mortals to cease foreseeing doom.
> *Chorus*: What cure did you provide them with against that sickness?
> *Prometheus*: I placed in them blind hopes.

The true Promethean spirit of science means to liberate man by giving him knowledge and some measure of dominion over the physical environment. But at another level, and in a new age, it also constructs the mythology of scientific materialism, guided by the corrective devices of the scientific method, addressed with precise and deliberately affective appeal to the deepest needs of human nature, and kept strong by the blind hopes that the journey on which we are now embarked will be farther and better than the one just completed.

Glossary
Notes
Index

Glossary

For the convenience of the reader I have prepared the following glossary of some of the terms used in this book that may be unfamiliar because they are technical or that, because of their importance, deserve a more than usually precise definition.

Adaptation. In biology, a particular anatomical structure, physiological process, or behavior that improves an organism's fitness to survive and reproduce. Also, the evolutionary process that leads to the acquisition of such a trait.

Aggression. Any physical act or threat of action by one individual that reduces the freedom or genetic fitness of another.

Altruism. Self-destructive behavior performed for the benefit of others. Altruism may be entirely rational, or automatic and unconscious, or conscious but guided by innate emotional responses.

Asexual reproduction. A form of reproduction, such as spore formation, budding, or simple cell division, that does not involve the fusion of sex cells.

Autocatalysis. The process in which the products of a reaction serve as catalysts, that is, they speed up the rate of the same reaction that produced them and cause it to accelerate.

Band. The term often applied to groups of hunter-gatherers.

Behavioral biology. The scientific study of all aspects of behavior, including neurophysiology (study of the nervous system), ethology (study of whole patterns of behavior), and sociobiology (study of the biological basis of social behavior and organization).

Budding. A form of asexual reproduction in which a more or less complete new organism simply grows from the body of the parent organism.

Carnivore. A creature that eats fresh meat.

Catalysis. The process by which a substance accelerates a reaction without itself being consumed in the overall course of the reaction.

Chromosome. A complex, often spherical or rod-shaped structure, found in the nucleus of cells and bearing part of the genetic information (genes) of the organism.

Cortex. In human anatomy, the outer layer of nervous tissue of the brain, the "gray matter" that contains the centers of consciousness and rational thought.

Darwinism. The theory of evolution by natural selection as argued by Charles Darwin (especially, in *The Origin of Species*, 1859). It holds that the genetic compositions of populations change through time—and thus evolve—first because individual members of the population vary among themselves in their hereditary material, and second because those endowed with the properties best fitting them for survival and reproduction will be disproportionately represented in later generations. This mode of evolution is viewed by modern biologists as the only one that operates beyond and above the mere statistical fluctuation of genetic types within populations.

Demography. The rate of growth and the age structure of populations, and the processes that determine these properties; also the scientific study of the properties.

Density dependence. An increase or decrease in the influence that some factor, such as disease or territorial behavior, exercises on the rate of population growth as a result of an increase in the density of the population.

Determinism. Loosely employed to designate any form of constraint on the development of an anatomical organ, physiological process, or behavior. Genetic determinism means some degree of constraint that is based on the possession of a particular set of genes.

Developmental landscape. A metaphor used to resolve the nature-nurture controversy. The development of a trait is compared to the passage of a ball rolling down a genetically fixed landscape, in which it comes periodically to divided channels and rolls into one or the other branches according to its momentum and the relative accessibility of the branches.

DNA (deoxyribonucleic acid). The fundamental hereditary material of all organisms. The genes are composed of the functional segments of DNA molecules.

Dominance system. In sociobiology, the set of relationships within a group of animals or men, often established and maintained by some form of aggression or coercion, in which one individual has precedence over all others in eating, mating, etc., a second individual has precedence over the remaining members of the group, and so on down a dominance hierarchy or "pecking order." Dominance orders are simple and strict in chickens but complex and subtle in human beings.

Drive. A term used loosely to describe the tendency of an animal to seek out an object, such as a mate, an item of food, or a nesting site, and to perform an appropriate response toward it.

Environmentalism. In the study of behavior, the belief that experience with the environment mostly or entirely determines the development of behavioral patterns.

Estrus. The period of heat, or maximum sexual receptivity, in the female. Under ordinary conditions the estrus is also the time of release of the female's eggs from the ovaries.

Ethology. The study of whole patterns of animal behavior in natural environments, with emphasis on analyzing adaptation and evolution of the patterns.

Evolution. Any gradual change. Organic evolution, often referred to as evolution for short, is any genetic change in a population of organisms from generation to generation.

Evolutionary biology. All of the branches of biology, including ecology, taxonomy, population biology, ethology, and sociobiology, that study the evolutionary process and the characteristics of whole populations and communities of organisms.

Fitness. See genetic fitness.

Gamete. A sex cell: an egg or a sperm.

Gene. A basic unit of heredity, a portion of the giant DNA molecule that affects the development of any trait at the most elementary biochemical level. The term gene is often applied more precisely to the cistron, the section of DNA that carries the code for the formation of a particular portion of a protein molecule.

Gene pool. All of the genes in an entire population of organisms.

Genetic. Hereditary; refers to variation in traits that is based at least in part on differences in genes.

Genetic fitness. The contribution to the next generation of one genetically distinct kind of organism relative to the contributions of other genetically different kinds belonging to the same population. By definition, those kinds with higher genetic fitness eventually come to prevail in the population; the process is called evolution by natural selection.

Genetics. The scientific study of heredity.

Genus. A group of similar, related species.

Gonad. An organ that produces sex cells; ordinarily, either an ovary (female gonad) or testis (male gonad).

Group selection. Any process, such as competition, the effects of disease, or the ability to reproduce, that results in one group of individuals leaving more descendants than another group. The "group" is loosely defined in theory: it can be a set of kin (usually more extended than merely parents and offspring; see kin selection), or part or all of a tribe or larger social group. Contrast with individual selection.

Haplodiploidy. The means of sex determination, such as that found in ants and other hymenopterous insects, in which males come from unfertilized eggs (hence they are haploid, having only one set of chromosomes) and females from fertilized eggs (making them diploid, or the possessors of two sets of chromosomes).

Hermaphroditism. The coexistence of both female and male sex organs in the same organism.

Homology. A similarity between anatomical structures, physiological processes, or behavioral patterns in two or more species due to the possession of a common ancestor and hence the possession of at least some genes that are identical by common descent.

Homozygous. Each ordinary cell in the body has two chromosomes of a kind; when the genes located at a given site on one of these chromosome pairs are identical to each another, the organism is said to be homozygous for that particular chromosome site.

Human nature. In the broader sense, the full set of innate behavioral predispositions that characterize the human species; and in the narrower sense, those predispositions that affect social behavior.

Hymenoptera. The insect order that contains all bees, wasps, and ants.

Hypergamy. The female practice of acquiring a mate of equal or higher social rank.

Hypertrophy. The extreme development of a preexisting structure. The elephant's tusk, for example, represents the hypertrophic enlargement and change in shape through evolution of a tooth that originally had an ordinary form. In this book it is suggested that most kinds of human social behavior are hypertrophic forms of original, simpler responses that were of more direct adaptive advantage in hunter-gatherer and primitively agricultural societies.

Hypothesis. A proposition that can be tested and is subject to possible disproof by further observation and experimentation. By the usual canons of scientific evidence, it is difficult if not impossible to prove a hypothesis with finality, but one can be tested so thoroughly and rigorously as to be transformed eventually into accepted fact—but never dogma. See theory.

Individual selection. Natural selection favoring the individual and its direct descendants. Contrast with group selection and kin selection.

Innate. Same as genetic: referring to variation based at least in part on differences in genes.

Instinct. Behavior that is relatively stereotyped, more complex than simple reflexes such as salivation and eye blinking, and usually directed at particular objects in the environment. Learning may or may not be involved in the development of instinctive behavior; the important point is that the behavior develops toward a comparatively narrow, predictable end product.

Because of its vagueness the term "instinct" is seldom used in technical scientific literature anymore, but it is so thoroughly entrenched in the English language—and useful as an occasional shorthand expression—that attempts at a precise definition are justified.

Kin selection. The increase of certain genes over others in a population as a result of one or more individuals favoring the survival and reproduction of relatives who therefore are likely to possess the same genes by common descent. Kin selection is one way in which altruistic behavior can evolve as a biological trait. Although kin are defined so as to include offspring, the term kin selection is ordinarily used only if at least some other relatives, such as brothers, sisters, or parents, are also affected. Contrast with individual selection.

Lamarckism. The theory, expounded by Jean Baptiste de Lamarck in 1809, that species evolve through physical and behavioral characteristics acquired by organisms during their lifetime and transmitted directly to their offspring. Lamarckism proved wrong as the explanation for biological evolution and was superseded by Darwinism, or evolution by natural selection.

Learning rule. A predisposition to learn one alternative behavior as opposed to another, even when both are taught with equal intensity. An example of a learning rule is the development of handedness: persons who are genetically right-handed can be trained to be left-handed only with difficulty, whereas the reverse is true of genetically left-handed persons.

Limbic system. A group of structures and regions in the deeper part of the forebrain that are interconnected and participate strongly in emotion, motivation, and reinforcement of learning. The principal parts include the hypothalamus, rhinencephalon (nosebrain), and hippocampus.

Mammal. Any animal of the class Mammalia (including man), characterized by the production of milk by the female mammary glands and the possession of hair for body covering.

Maturation. The automatic development of a pattern of behavior which becomes increasingly complex or precise as the animal matures. Unlike learning, the development does not require experience to occur.

Mutation. In the broad sense, any discontinuous change in the genetic constitution of an organism. A mutation can consist of a change in the chemical structure of a gene (segment of DNA) or in the structure or number of entire chromosomes.

Natural selection. The differential contribution of offspring to the next generation by various genetic types belonging to the same population. This mechanism of evolution was suggested by Charles Darwin and is thus also called Darwinism. It has been supported and greatly strengthened by the findings of modern genetics.

Neurobiology. The scientific study of the anatomy (neuroanatomy) and physiology (neurophysiology) of the nervous system.

Neuron. A nerve cell; the basic unit of the nervous system.

Neurophysiology. See neurobiology.

Nucleus. The central body of the cell, containing the hereditary material of the organism. (Genes are carried on structures within the nucleus called chromosomes.)

Ontogeny. The development of a single organism throughout its lifetime (contrast with phylogeny).

Phylogeny. The evolutionary history of a particular group of organisms; also, the "family tree" that shows which species gave rise to others (contrast with ontogeny).

Physiology. The scientific study of the functions of living organisms and the individual organs, tissues, and cells of which they are composed.

Polygamy. The possession of multiple mates by an individual, either multiple females by a male (polygyny) or multiple males by a female (polyandry).

Polygyny. The possession of two or more mates by a male.

Population. Any group of organisms capable of interbreeding for the most part and coexisting at the same time and in the same place.

Prepared learning. An innate predisposition to learn one thing as opposed to another, even when the intensity of training is made equal for both. For example, a person who is genetically right-handed is prepared to learn use of the right hand and deterred from learning to use the left hand, or can be induced only by special effort to do so.

Primate. A member of the order Primates, such as a lemur, monkey, ape, or man.

Reciprocal altruism. The trading of altruistic acts by individuals at different times. For example, one person saves a drowning person in exchange for the promise (or at least the reasonable expectation) that the altruistic act will be repaid if circumstances are ever reversed.

Scientific materialism. The view that all phenomena in the universe, including the human mind, have a material basis, are subject to the same physical laws, and can be most deeply understood by scientific analysis.

Selection. See natural selection.

Sex ratio. The ratio of males to females (for example, two males to one female) in a population or society.

Social insect. One of the kinds of insect that form colonies with reproductive castes and worker castes; in particular, the termites, ants, social bees, and social wasps.

Sociality. The combined properties and processes of social existence.

Society. A group of individuals belonging to the same species and organized in a cooperative manner. The principal criterion for applying the term "society" is the existence of reciprocal communication of a cooperative nature that extends beyond mere sexual activity.

Sociobiology. The scientific study of the biological basis of all forms of social behavior in all kinds of organisms, including man.

Species. A population or set of populations of closely related and similar organisms, which ordinarily breed freely among themselves and not with members of other populations.

Taxonomy. The science and art of the classification of organisms.

Territory. A fixed area from which an organism or group of organisms excludes other members of the same species by aggressive behavior or display.

Theory. A set of broad propositions about some process in nature, such as the mode of evolution or the history of the earth's continents, that lead to the creation of conjectures—"hypotheses" —about specific phenomena that can be tested. A theory is regarded as truthful if it stimulates the invention of new hypotheses, if the hypotheses stand up under testing, and if as a result the explanations made possible by the theory are more effective and satisfying in explaining some part of reality than the explanations pressed by rival theories.

Zoology. The scientific study of animals.

Notes

Chapter 1: Dilemma

1 Hans Küng, *On Being a Christian*, translated by Edward Quinn (Doubleday, New York, 1976).

2 In order to encompass such views the expression "new naturalism" has been used by David Mathews, in David Mathews et al., *The Changing Agenda of Higher Education* (U.S. Government Printing Office, Washington, D.C., 1977); and in "The American Achievement in Education: A Self-education Society under the Tutelage of Nature," *Frontiers of Knowledge* (Doubleday, New York, in press).

2 Steven Weinberg, "The Forces of Nature," *Bulletin of the American Academy of Arts and Sciences* 29(4): 13–29 (1976).

3 W. B. Yeats, "The coming of wisdom with time" (1910), in Peter Allt and R. K. Alspach, eds., *The Variorum Edition of the Poems of W. B. Yeats* (Macmillan Co., New York, 1957. Reprinted by permission of M. B. Yeats, Miss Anne Yeats, Macmillan Publishing Company of New York, and The Macmillan Company of London & Basingstoke).

3 Alain Peyrefitte, *The Chinese: Portrait of a People*, translated from the French by Graham Webb (Bobbs-Merrill, New York, 1977).

PAGE

4 Gunther S. Stent, *The Coming of the Golden Age: A View of the End of Progress* (Natural History Press, Garden City, Long Island, New York, 1969).

5 The idea of the genetic evolution of moral predispositions by natural selection has a long but relatively ineffectual history. Charles Darwin raised the possibility in *The Descent of Man and Selection in Relation to Sex* (London, 1971), and he was firm in disputing the view held by John Stuart Mill and Alfred Russel Wallace that the mind has been freed from natural selection. He felt that if the human mentality were excepted, the basic theory of evolution by natural selection would be gravely threatened; to Wallace, the co-discoverer of natural selection, he wrote in 1869, "I hope you have not murdered too completely your own and my child" (*More Letters of Charles Darwin*, edited by Francis Darwin, D. Appleton, New York, vol. 2, p. 39, 1903). Darwin's thoughts on this subject had run deep. In his unpublished notes of July 1838 he took the optimistic view that an understanding of evolution would lead to a stronger morality: "Two classes of moralists: one says our rule of life is what *will* produce the greatest happiness. — The other says we have a moral sense. — But my view unites both and shows them to be almost identical and what *has* produced the greatest good or rather what was necessary for good at all *is* the instinctive moral sense" (pp. 242–243 in Howard E. Gruber, *Darwin on Man: A Psychological Study of Scientific Creativity*, together with Darwin's early and unpublished notebooks transcribed and annotated by Paul H. Barrett, E. P. Dutton, New York, 1974).

Herbert Spencer, the most ambitious of the nineteenth century evolutionists, argued the necessity of a non-Kantian, rationalistic approach to ethics (*Principles of Ethics*, New York, 1896). He believed that the human nervous system has been modified through thousands of generations to create certain innate faculties of moral intuition, consisting of emotions responding to right and wrong conduct, but that human nature can be molded by "the rigorous maintenance of the conditions of harmonious social co-operation" (*An Autobiography*, D. Appleton, New York, vol. 2, p. 8, 1904).

5   In *The Influence of Darwin on Philosophy* (P. Smith, New York,
1910), John Dewey concluded that evolutionary theory, and
specifically Darwinism, provides the means to fashion a scientific
ethics; but later, in *Human Nature and Conduct* (Holt, New
York, 1922), he conceded that specific ethical premises are culturally acquired.

More recently, Antony Flew in *Evolutionary Ethics* (Macmillan, London, 1967), attempting to refute Wittgenstein's claim that
evolutionary theory is irrelevant to philosophy, extends the idea
that ethical behavior has evolved and hence is subject to empirical
evaluation. In *Sociobiology: The New Synthesis* (The Belknap
Press of Harvard University Press, Cambridge, Mass., 1975) and
"The Social Instinct," *Bulletin of the American Academy of Arts
and Sciences* 30(1): 11–25 (1976), I relate the genetic evolution
of ethical predisposition to specific principles in population biology. Gunther Stent, in *The Hastings Center Report* 6(6): 32–40
(1976), discusses the promise and limitations of a "structuralist
ethics." The theme is carried still further by George E. Pugh in
The Biological Origin of Human Values (Basic Books, New York,
1977), an important work combining ideas from mathematical
control theory and biology.

In broader terms, Konrad Lorenz has pioneered in the development of the conception of cognition and thought as evolutionary
products of a structured brain. His most recent views are contained in *Behind the Mirror: A Search for a Natural History of
Human Knowledge* (translated from the German by Ronald Tay-
lor; Harcourt Brace Jovanovich, New York, 1977). A favorable
critique of Lorenz's contributions, with original extensions and a
historical review, is provided by Donald T. Campbell in "Evolutionary epistemology," in Paul Schilpp, ed., *The Philosophy of
Karl Popper* (Open Court, La Salle, Illinois, 1974, pp. 415–463).
See also the more popularized and personal account by Richard
I. Evans, *Konrad Lorenz: The Man and His Ideas* (Harcourt
Brace Jovanovich, New York, 1975).

7   The idea of sociobiology as the antidiscipline of the social sciences
was presented in my article "Biology and the Social Sciences,"

PAGE

7   *Daedalus* 106(4): 127–140 (1977). Portions of the article have been reproduced here by permission of the editors of *Daedalus*, the Journal of the American Academy of Arts and Sciences.

10   The classic statement of the discontinuity between science and the humanities was made by Charles P. Snow in *The Two Cultures and the Scientific Revolution* (Cambridge University Press, Cambridge, 1959).

10   Theodore Roszak, "The Monster and the Titan: Science, Knowledge, and Gnosis," *Daedalus* 103(3): 17–32 (1974).

11   Ernst Mach, *The Science of Mechanics*, ninth edition (Open Court, LaSalle, Illinois, 1942).

Chapter 2: Heredity

15   Howard E. Evans, *Life on a Little-Known Planet* (Dutton, New York, 1968).

16   An introduction to social organisms and the discipline of sociobiology is given in Wilson, *Sociobiology*.

16   An excellent review of modern ethology, with a detailed chapter on human fixed-action patterns, is provided by Irenäus Eibl-Eibesfeldt in *Ethology: The Biology of Behavior*, second edition (Holt, Rinehart and Winston, New York, 1977). The most original and authoritative synthesis of ethology and comparative psychology is contained in Robert A. Hinde's *Animal Behavior*, second edition (McGraw-Hill, New York, 1970).

17   J. J. Rousseau, *Essai sur l'origine des langues*, Oeuvres Posthumes, vol. 2 (London, 1783); quoted by Claude Lévi-Strauss in *La Pensée Sauvage* (Plon, Paris, 1964).

17   Robert Nozick, *Anarchy, State, and Utopia* (Basic Books, New York, 1974).

18   The machine-like qualities of human information processing are explained in Allen Newell and Herbert A. Simon, *Human Problem Solving* (Prentice-Hall, Englewood Cliffs, New Jersey, 1972), and George Boolos and Richard Jeffrey, *Computability and Logic* (Cambridge University Press, Cambridge, 1974).

19   The inheritance of eye color is discussed in Curt Stein, *Principles of Human Genetics*, 3d ed. (W. H. Freeman, San Francisco, 1973).

Notes

PAGE

20   R. D. Alexander, J. L. Hoogland, R. D. Howard, K. M. Noonan, and P. W. Sherman, "Sexual Dimorphisms and Breeding Systems in Pinnipeds, Ungulates, Primates, and Humans," in N. A. Chagnon and W. G. Irons, eds., *Evolutionary Biology and Human Social Organization* (Duxbury Press, Scituate, Mass., in press).

21   The evidence of the destructive long-term effects of abnormal experience during early development is reviewed by Ronald P. Rohner, *They Love Me, They Love Me Not* (HRAF Press, New Haven, Conn., 1975), and T. G. R. Bower, *A Primer of Infant Development* (W. H. Freeman, San Francisco, 1977).

21   Theodosius Dobzhansky, "Anthropology and the Natural Sciences — The Problem of Human Evolution," *Current Anthropology* 4: 138, 146–148 (1963).

21   George P. Murdock, "The Common Denominator of Culture," in Ralph Linton, ed., *The Science of Man in the World Crisis* (Columbia University Press, New York, 1945), pp. 124–142.

23   Robin Fox, "The Cultural Animal," in J. F. Eisenberg and W. S. Dillon, eds., *Man and Beast: Comparative Social Behavior* (Smithsonian Institution Press, Washington, D.C., 1971), pp. 273–296.

25   Mary-Claire King and Allan C. Wilson, "Evolution at two levels in humans and chimpanzees," *Science* 188: 107–116 (1975).

25   The ability of chimpanzees to learn language has been reviewed in David Premack, "Language and Intelligence in Ape and Man," *American Scientist* 64(6): 674–683 (1976); and Carl Sagan, *The Dragons of Eden* (Random House, New York, 1977).

26   The early evolution of the human larynx and capacity for language has been analyzed in Jan Wind, "Phylogeny of the Human Vocal Tract," *Annals of the New York Academy of Sciences* 280: 612–630 (1976); and Philip Lieberman, "The Phylogeny of Language," in T. A. Sebeok, ed., *How Animals Communicate* (Indiana University Press, Bloomington, 1977), pp. 3–25.

26   Leslie A. White, *The Science of Culture: A Study of Man and Civilization* (Farrar, Straus and Giroux, New York, 1949).

26   Gordon G. Gallup, "Self-Recognition in Primates: A Comparative Approach to the Bidirectional Properties of Consciousness," *American Psychologist* 32(5): 329–338 (1977).

27   David Premack, "Language and intelligence."

PAGE

28 The early stages of territorial aggression in the Gombe chimpanzee population are mentioned in Glenn E. King, "Socioterritorial Units among Carnivores and Early Hominids," *Journal of Anthropological Research* 31(1): 69–87 (1975). Other details are given in Jane Lancaster, "Carrying and Sharing in Human Evolution," *Human Nature* 1(2): 82–89 (1978); while a more theoretical discussion of the causes of the phenomenon is found in Richard W. Wrangham, "On the Evolution of Ape Social Systems" in Irven DeVore, ed., *Sociobiology and the Social Sciences* (Aldine, Chicago, in press).

28 Richard B. Lee, "What Hunters Do for a Living, or, How to Make Out on Scarce Resources," in R. B. Lee and Irven DeVore, eds., *Man the Hunter* (Aldine, Chicago, 1968), pp. 30–48.

28 Chimpanzee hunting behavior is described in Geza Teleki, *The Predatory Behavior of Wild Chimpanzees* (Bucknell University Press, Lewisburg, Pa., 1973).

30 Jane van Lawick-Goodall (Jane Goodall), "The Behaviour of Free-Living Chimpanzees in the Gombe Stream Reserve," *Animal Behaviour Monographs* 1(3): 161–311 (1968); "Mother-Offspring Relationships in Free-Ranging Chimpanzees," in Desmond Morris, ed., *Primate Ethology* (Aldine, Chicago, 1969), pp. 364–436; "Tool-using in Primates and Other Vertebrates," *Advances in the Study of Behavior* 3: 195–249 (1970).

31 Jorge Sabater-Pí, "An Elementary Industry of the Chimpanzees in the Okorobikó Mountains, Rio Muni (Republic of Equatorial Africa), West Africa," *Primates* 15(4): 351–364 (1974).

33 Recent critiques of the modern version of natural selection theory are given in Anthony Ferguson, "Can Evolutionary Theory Predict?" *American Naturalist* 110: 1101–1104 (1976); G. Ledyard Stebbins, "In Defense of Evolution: Tautology or Theory?" *American Naturalist* 111: 386–390 (1977); Theodosius Dobzhansky, Francisco J. Ayala, G. Ledyard Stebbins, and James W. Valentine, *Evolution* (W. H. Freeman, San Francisco, 1977); and George F. Oster and Edward O. Wilson, "A Critique of Optimization Theory in Evolutionary Biology," in *Caste and Ecology in the Social Insects* (Princeton University Press, Princeton, N.J., 1978).

PAGE

36 Joseph Shepher, "Mate Selection among Second-Generation Kibbutz Adolescents and Adults: Incest Avoidance and Negative Im-
printing," *Archives of Sexual Behavior* 1(4): 293–307 (1971).
The possibility of automatic aversion based on early domestic
intimacy was first suggested by Edward Westermarck in 1891.

36 All three of the principal explanations of the incest taboo were
first formulated in the late nineteenth century, during the flourish-
ing period of evolutionism in anthropology: the family-integrity
hypothesis by Carl N. Starcke (1889), the alliance hypothesis by
Edward Tylor (1889), and the inbreeding-depression hypothesis
by Lewis Henry Morgan (1877). The history of the subject has
been reviewed by Marvin Harris in *The Rise of Anthropological
Theory* (Thomas Y. Crowell, New York, 1968). A thorough
cross-cultural review, which considers all of the competing hypotheses and gives the edge to the biological explanation, has been
provided in Melvin Ember, "On the Origin and Extension of the
Incest Taboo," *Behavior Science Research* (Human Relations
Area Files, New Haven, Connecticut) 10: 249–281 (1975).

37 For general accounts of recessive genes and the deleterious effects
of inbreeding in human beings, see Curt Stern, *Principles of Human Genetics*, 3d ed. (W. H. Freeman, San Francisco, 1973); and
L. L. Cavalli-Sforza and W. F. Bodmer, *The Genetics of Human
Populations* (W. H. Freeman, San Francisco, 1971). The estimate
of lethal genes in human populations is in N. E. Morton, J. F.
Crow, and H. J. Muller, "An Estimate of the Mutational Damage
in Man from Data on Consanguineous Marriages," *Proceedings of
the National Academy of Sciences, U.S.A.* 42: 855–863 (1956).
The research on Czechoslovakian children born from incestuous
unions was conducted by Eva Seemanova, as reported in *Time*,
October 9, 1972.

39 R. L. Trivers and D. E. Willard, "Natural Selection of Parental
Ability to Vary the Sex Ratio of Offspring," *Science* 179: 90–92
(1973).

40 Mildred Dickeman, "Female Infanticide and the Reproductive
Strategies of Stratified Human Societies: A Preliminary Model,"
in Napoleon A. Chagnon and William G. Irons, eds., *Evolutionary*

PAGE

40   *Biology and Human Social Organization* (Duxbury Press, Scituate, Mass., 1978).

41   Richard H. Wills, *The Institutionalized Severely Retarded* (Charles C Thomas, Springfield, Ill., 1973).

43   Reviews of human behavioral genetics are provided in G. E. McClearn and J. C. DeFries, *Introduction to Behavioral Genetics* (W. H. Freeman, San Francisco, 1973); and Lee Ehrman and P. A. Parsons, *The Genetics of Behavior* (Sinauer Associates, Sunderland, Mass., 1976).

43   H. A. Witkin et al., "Criminality in XYY and XXY Men," *Science* 193: 547–555 (1976).

44   The Lesch-Nyhan and Turner's syndromes are described in J. C. DeFries, S. G. Vandenberg, and G. E. McClearn, "Genetics of Specific Cognitive Abilities," *Annual Review of Genetics* 10: 179–207 (1976); and C. R. Lake and M. G. Ziegler, "Lesch-Nyhan Syndrome: Low Dopamine-β-Hydroxylase Activity and Diminished Response to Stress and Posture," *Science* 196: 905–906 (1977).

45   The twin-analysis method is described more fully in G. E. McClearn and J. C. DeFries, *Introduction to Behavioral Genetics.* Special studies of interest are L. L. Heston and J. Shields, "Homosexuality in Twins: a Family Study and a Registry Study," *Archives of General Psychiatry* 18: 149–160 (1968); and N. G. Martin, L. J. Eaves, and H. J. Eysenck, "Genetical, Environmental and Personality Factors in Influencing the Age of First Sexual Intercourse in Twins," *Journal of Biosocial Science* 9(1): 91–97 (1977). In addition, Sandra Scarr and Richard A. Weinberg have provided important new evidence of the inheritance of intelligence and personality traits based on comparisons of children raised by biological as opposed to adoptive parents ("Attitudes, Interests, and IQ," *Human Nature* 1(4): 29–36, 1978). Although substantial genetic variation occurs among families within the same population, Scarr and Weinberg could find no evidence of average differences in IQ between Americans of African and European descent.

46   J. C. Loehlin and R. C. Nichols, *Heredity, Environment, and Personality* (University of Texas Press, Austin, 1976).

PAGE

47 V. A. McKusick and F. H. Ruddle, "The status of the gene map of the human chromosome," *Science* 196: 390–405 (1977).

47 See for example Joan Arehart-Treichel, "Enkephalins: More than Just Pain Killers," *Science News* 112(4): 59, 62 (1977).

48 For a fuller analysis of the nature of geographic variation, see Ed-O. Wilson and William L. Brown, "The subspecies concept and its taxonomic application," *Systematic Zoology* 2(3): 97–111 (1953).

48 Daniel G. Freedman, *Human Infancy: An Evolutionary Perspective* (Lawrence Erlbaum, Hillsdale, N.J., 1974).

49 Nova Green, "An Exploratory Study of Aggression and Spacing in Two Preschool Nurseries: Chinese-American and European-American" (master's thesis, University of Chicago, 1969).

50 Marvin Bressler, "Sociology, Biology and Ideology," in David Glass, ed., *Genetics* (Rockefeller University Press, New York, 1968), pp. 178–210.

Chapter 3: Development

54 Descriptions of the visual neurons are provided, with a penetrating philosophical discussion, in Gunther S. Stent, "Limits to the Scientific Understanding of Man," *Science* 187: 1052–1057 (1975); and by one of the major investigators of the subject, David H. Hubel, in "Vision and the brain," *Bulletin of the American Academy of Arts and Sciences* 31: 17–28 (1978); the auditory system is described by Harry J. Jerison, "Fossil Evidence of the Evolution of the Human Brain," *Annual Review of Anthropology* 4: 27–58 (1975).

55 For a thorough philosophical discussion of determinism, including its possible meaning in psychology, see Bernard Berofsky's *Determinism* (Princeton University Press, Princeton, N.J., 1971).

55 The mosquito example and other case histories of stereotyped behavior are found in Thomas Eisner and Edward O. Wilson, eds., *Animal Behavior* (W. H. Freeman, San Francisco, 1976).

57 The evidence for inheritance of handedness is given in Curt Stern, *Principles of Human Genetics*. However, many of the important data have been reanalyzed and their significance challenged by

PAGE

57 Robert L. Collins ("The Sound of One Paw Clapping: An Inquiry into the Origin of Left-Handedness," in Gardner Lindzey and Delbert D. Thiessen, eds., *Contributions to Behavior-Genetic Analysis: The Mouse as a Prototype* (Appleton-Century-Crofts, New York, 1970). Collins prefers the explanation that handedness is due either to unknown biological influences on the fetus or to the inheritance of a learning rule — a strong predisposition to choose one side or the other at an early age, with the side picked depending on chance or culture. Teng's Chinese studies, cited in the note below, appear to favor prenatal determination as opposed to a learning rule. This general explanation (which includes the purely genetic hypothesis) is also supported by the fact that left-handed persons have constituted a small minority, perhaps on the order of ten percent, since prehistory; see Curtis Hardyk and Lewis F. Petrinovich, "Left-handedness," *Psychological Bulletin* 84: 385–404 (1977).

57 Evelyn Lee Teng, Pen–hua Lee, K. Yang, and P. C. Chang, "Handedness in Chinese Populations: Biological, Social, and Pathological Factors," *Science* 193: 1146–1150 (1976).

58 T. S. Szasz, *The Myth of Mental Illness: Foundations of a Theory of Personal Conduct*, revised edition (Harper & Row, New York, 1974). R. D. Laing and A. Esterson, *Sanity, Madness and the Family* (Tavistock, London, 1964).

59 Account of lectures by Seymour S. Kety and Steven Matthysse, "Genetic Aspects of Schizophrenia," in Bernard D. Davis and Patricia Flaherty, eds., *Human Diversity: Its Causes and Social Significance* (Ballinger, Cambridge, Mass., 1976), pp. 108–115.

59 Jane M. Murphy, "Psychiatric Labeling in Cross-Cultural Perspective," *Science* 191: 1019–1028 (1976).

59 The research by Philip Seeman and Tyrone Lee on dopamine receptors is reported in *Science News* 112: 342 (1977).

59 The traits of the schizophrenogenic family and other factors influencing schizophrenia are well described in Roger Brown and Richard J. Herrnstein, *Psychology* (Little, Brown, Boston, Mass., 1975).

60 See especially *Evolution and Modification of Behavior*, by Konrad

PAGE

60 Lorenz (Phoenix Books, University of Chicago Press, Chicago, 1965); Robert A. Hinde, *Animal Behavior*; and B. F. Skinner, "The phylogeny and ontogeny of behavior," *Science* 153: 1205–1213 (1966).

60 C. H. Waddington, *The Strategy of the Genes: A Discussion of Aspects of Theoretical Biology* (George Allen and Unwin, London, 1957).

61 Paul Ekman and Wallace V. Friesen, *Unmasking the Face* (Prentice-Hall, Englewood Cliffs, N.J., 1975); and Paul Ekman, "Darwin and Cross-Cultural Studies of Facial Expression," in Paul Ekman, ed., *Darwin and Facial Expression: A Century of Research in Review* (Academic Press, New York, 1973).

61 Irenäus Eibl-Eibesfeldt, *Ethology: The Biology of Behavior*, 2d ed. (Holt, Rinehart and Winston, New York, 1977).

62 The information on smiling in blind children is reviewed in Eibl-Eibesfeldt, *Ethology*.

62 Melvin J. Konner, "Aspects of the Developmental Ethology of a Foraging People," in N. G. Blurton Jones, ed., *Ethological Studies of Child Behaviour* (Cambridge University Press, 1972), pp. 285–304; and quoted in Joel Greenberg, "The Brain and Emotions," *Science News* 112: 74–75 (1977).

62 The evidences of canalized development of smiling in infants with normal vision must be evaluated with caution. Recently, the British psychologists Andrew N. Meltzoff and M. Keith Moore showed that infants as young as two weeks can imitate a variety of facial expressions and manual gestures performed by adults close to them ("Imitation of Facial and Manual Gestures by Human Neonates," *Science* 198: 75–78, 1977). The evidence from blind and blind-deaf children remains unchallenged, however.

63 The necessity of programmed language acquisition is discussed in G. A. Miller, E. Galanter, and K. H. Pribram, *Plans and the Structure of Behavior* (Henry Holt, New York, 1960). Roger Brown describes the early ontogeny of language in *A First Language: The Early Stages* (Harvard University Press, Cambridge, Mass., 1973).

PAGE

64 Skinner, B. F., *The Behavior of Organisms* (Appleton, New York, 1938).

65 The concept of the constraint of learning as a biological adaptation is discussed fully in Martin E. P. Seligman and Joanne L. Hager, eds., *Biological Boundaries of Learning* (Prentice-Hall, Englewood Cliffs, N.J., 1972).

65 The examples of prepared learning in animals are provided in Seligman and Hager, eds., *Biological Boundaries*; J. S. Rosenblatt, "Learning in Newborn Kittens," *Scientific American* 227(6): 18–25 (1972); Sara J. Shettleworth, "Constraints on Learning," *Advances in the Study of Behavior* 4: 1–68 (1972), and "Conditioning of Domestic Chicks to Visual and Auditory Stimuli," in Seligman and Hager, eds., *Biological Boundaries*, pp. 228–236; and Stephen T. Emlen, "The Stellar-Orientation System of a Migratory Bird," *Scientific American* 233(2): 102–111 (1975).

66 Jean Piaget, *Genetic Epistemology*, translated from the French by Eleanor Duckworth (Columbia University Press, New York, 1970). See also *The Origins of Intellect: Piaget's Theory*, 2d ed., by John L. Phillips, Jr. (W. H. Freeman, San Francisco, 1975).

67 John Bowlby, *Attachment* (Basic Books, New York, 1969); *Separation: Anxiety and Anger* (Basic Books, New York, 1973).

67 Lawrence Kohlberg, "Stage and Sequence: The Cognitive-Descriptive Approach to Socialization," in D. A. Goslin, ed., *Handbook of Socialization Theory and Research* (Rand-McNally, Chicago, Ill., 1969), pp. 347–480.

67 The comparison of heritability in various categories of ability and personality traits is provided in S. G. Vandenberg, "Heredity Factors in Normal Personality Traits (as Measured by Inventories)," *Recent Advances in Biological Psychiatry* 9: 65–104 (1967); and J. C. Loehlin and R. C. Nichols, *Heredity, Environment, and Personality* (University of Texas Press, Austin, 1976). The idea of the adaptive significance of the differences is due to D. G. Freedman, *Human Infancy: An Evolutionary Perspective* (Lawrence Erlbaum Associates, Hillsdale, N.J., 1974).

68 The significance of phobias is discussed in M. E. P. Seligman, "Phobias and Preparedness," in Seligman and Hager, eds., *Biological Boundaries*, pp. 451–462.

68   Lionel Tiger and Robin Fox, *The Imperial Animal* (Holt, Rinehart and Winston, New York, 1971).

70   Erik H. Erikson, *Identity: Youth and Crisis* (W. W. Norton, New York, 1968).

Chapter 4: Emergence

74   The description of the neurobiology of vision is based on the article by Gunther S. Stent, "Limits to the Scientific Understanding of Man," *Science* 187: 1052–1057 (1975).

74   Charles Sherrington, *Man on His Nature* (Cambridge University Press, Cambridge, 1940).

75   The concept of the brain schema or plan is reviewed in G. A. Miller, E. Galanter, and K. H. Pribram, *Plans and the Structure of Behavior* (Holt, Rinehart and Winston, New York, 1960); and Ulric Neisser, *Cognition and Reality* (W. H. Freeman, San Francisco, 1976).

76   Oliver Sacks, "The Nature of Consciousness," *Harper's* 251 (1507):5 (December 1975).

77   The complex relationships of brain, mind, individuality, determinism, free will, and fatalism have of course been central topics of philosophy for centuries and now also hold the attention of theoretical psychologists. The view presented here is both personal and greatly simplified. Especially useful works that explore the subject in greater detail include *The Concept of Mind* by Gilbert Ryle (Hutchinson, London, 1949); *The Concept of a Person, and Other Essays* by A. J. Ayer (St. Martin's Press, New York, 1963); and the historical review and anthology provided in Antony Flew, *Body, Mind, and Death* (Macmillan, New York, 1964).

78   The analysis of honeybee flight characteristics is presented in Karl von Frisch, *The Dance Language and Orientation of Bees*, translated from the German by L. Chadwick (Belknap Press of Harvard University Press, Cambridge, Mass., 1967); and George F. Oster and Edward O. Wilson, *Caste and Ecology in the Social Insects* (Princeton University Press, Princeton, N.J., 1978).

78   Several aspects of a more technical theory of the interaction of

PAGE

78 genetic and cultural evolution have been presented in L. L. Caval-li-Sforza and M. W. Feldman, "Models for Cultural Inheritance: I. Group Mean and within Group Variation," *Theoretical Population Biology* 4: 42–55 (1973); Robert Boyd and P. J. Richerson, "A Simple Dual Inheritance Model of the Conflict between Social and Biological Evolution," *Zygon* 11: 254–262 (1976); and W. H. Durham, "The Adaptive Significance of Cultural Behavior," *Human Ecology* 4: 89–121 (1976).

80 Lionel Trilling, *Beyond Culture: Essays on Literature and Learning* (Viking Press, New York, 1955).

80 Orlando Patterson, "Slavery," *Annual Review of Sociology* 3: 407–449 (1977); and "The Structural Origins of Slavery: A Critique of the Nieboer-Domar Hypothesis from a Comparative Perspective," *Annals of the New York Academy of Sciences* 292: 12–34 (1977).

83 Richard B. Lee, "What Hunters Do for a Living, or How to Make Out on Scarce Resources," in R. B. Lee and Irven DeVore, eds., *Man the Hunter* (Aldine, Chicago, 1968) pp. 30–48.

84 The parallels of early human social organization to that of four-footed carnivores have been explored in G. B. Schaller and G. R. Lowther, "The Relevance of Carnivore Behavior to the Study of Early Hominids," *Southwestern Journal of Anthropology* 25(4): 307–3441 (1969); and P. R. Thompson in "A Cross-Species Analysis of Carnivore, Primate, and Hominid Behavior," *Journal of Human Evolution* 4(2): 112–124 (1975).

84 The account of the autocatalysis model of human social evolution is from Wilson, *Sociobiology*, pp. 566–568. The archeological evidence of the ecology and food habits of the earliest human beings has been ably summarized by Glynn Isaac in "The Food-Sharing Behavior of Protohuman Hominids," *Scientific American* 238: 90–108 (April 1978).

85 The campsite talk of the !Kung is described in Richard B. Lee, "The !Kung Bushmen of Botswana," in M. G. Bicchieri, ed., *Hunters and Gatherers Today* (Holt, Rinehart and Winston, New York 1972), pp. 327–368.

85 Excellent descriptions of life in hunter-gatherer societies are given in John E. Pfeiffer, *The Emergence of Man* (Harper & Row,

PAGE

85   New York, 1969) and *The Emergence of Society* (McGraw-Hill, New York, 1977).

86   Robin Fox, "Alliance and Constraint: Sexual Selection in the Evolution of Human Kinship Systems," in B. G. Campbell, ed., *Sexual Selection and the Descent of Man 1871–1971* (Aldine, Chicago, 1972), pp. 282–331.

87   The estimates of the rates of the evolutionary increase in human brain size are based on all existing fossil data published to 1977 and were generously provided by Harry J. Jerison (personal communication).

89   Kent V. Flannery, "The cultural evolution of civilizations," *Annual Review of Ecology and Systematics* 3: 399–426 (1972).

90   Kent V. Flannery, "The cultural evolution of civilizations." (Copyright © 1972 by Annual Reviews, Inc. All rights reserved.)

91   Draper, Patricia, "!Kung Women: Contrasts in Sexual Egalitarianism in Foraging and Sedentary Contexts," in Rayna R. Reiter, ed., *Toward an Anthropology of Women* (Monthly Review Press, New York, 1975), pp. 77–109.

92   Erving Goffman, *Frame Analysis* (Harvard University Press, Cambridge, Mass., 1974).

93   Marvin Harris, *Cannibals and Kings: The Origins of Cultures* (Random House, New York, 1977).

94   The hypothesis of the cannibalistic origin of Aztec sacrifice is due to Michael Harner; see "The Enigma of Aztec Sacrifice," *Natural History* 84: 46–51 (April 1977). It has been challenged by other anthropologists, who doubt the evidence of the protein insufficiency in the Aztec diet. See, for example, Michael D. Coe, "Struggles of Human History," *Science* 199: 762–763 (1978); and "Demystification, Enriddlement, and Aztec Cannibalism: A Materialist Rejoinder to Harner," by Barbara J. Price, *American Ethnologist* 5: 98–115 (1978).

96   This account of the development of computer technology is based on Robert Jastrow's article, "Post-Human Intelligence," *Natural History* 84: 12–18 (June–July 1977). Note that the capacity cited is memory and does not necessarily encompass the little-understood, and possibly even more complex, processes of language formation and decision making.

Chapter 5: Aggression

99 Data on the frequency of wars are provided in Pitirim Sorokin, *Social and Cultural Dynamics* (Porter Sargent, Boston, 1957); see also Quincy Wright's classic, *A Study of War*, 2d ed. (University of Chicago Press, Chicago, 1965).

100 Elizabeth Marshall Thomas, *The Harmless People* (Alfred Knopf, New York, 1959).

100 !Kung San murder rate: based on a talk, "!Kung Bushman violence," by Richard B. Lee at the annual meeting of the American Anthropological Association, November 1969.

100 Robert K. Dentan, *The Semai: A Nonviolent People of Malaya* (Holt, Rinehart and Winston, New York, 1968).

101 A discussion of behavioral scales and other properties of aggressive behavior is presented in Wilson, *Sociobiology*, pp. 19–21, 242–297.

101 Sigmund Freud, "Why war," in *Collected Papers* (J. Strachey, ed.) vol. 5 (Basic Books, New York, 1959), pp. 273–287.

101 Konrad Lorenz, *On Aggression* (Harcourt, Brace & World, New York, 1966).

101 Erich Fromm, *The Anatomy of Human Destructiveness* (Holt, Rinehart and Winston, New York, 1973).

101 The diversity of kinds of aggressive behavior is reviewed in *Sociobiology*, pp. 242–255.

102 The example of rattlesnake aggression is from George W. Barlow, "Ethological Units of Behavior," in D. Ingle, ed., *The Central Nervous System and Fish Behavior* (University of Chicago Press, Chicago, 1968), pp. 217–232.

103 I first formulated this relationship between aggression and ecology in "'Competitive and Aggressive Behavior," in J. F. Eisenberg and W. Dillon, eds., *Man and Beast: Comparative Social Behavior* (Smithsonian Institution Press, Washington, D.C., 1971), pp. 183–217.

103 A more recent and accurate account of animal aggression is provided in Boyce Rensberger's *The Cult of the Wild* (Anchor Press, Doubleday, Garden City, New York, 1977).

103 Some of the account of animal aggression is taken from my article,

PAGE

103 "Human decency is animal," *New York Times Magazine*, 12 October 1975, pp. 38–50 (copyright © 1975 by the New York Times Company; reprinted by permission).

104 Hans Kruuk, *The Spotted Hyena: A Study of Predation and Social Behavior* (University of Chicago Press, Chicago, 1972).

105 R. G. Sipes, "War, Sports and Aggression: An Empirical Test of Two Rival Theories," *American Anthropologist* 75: 64–86 (1973); see an account of Sipes' more recent research in *Science News*, December 13, 1975, p. 375.

107 A summary of territorial conflict among hunter-gatherer bands is given in Glenn E. King, "Society and Territory in Human Evolution," *Journal of Human Evolution* 5: 323–332 (1976).

108 Rada Dyson-Hudson and Eric A. Smith, "Human Territoriality: An Ecological Assessment," in Napoleon Chagnon and William Irons, eds., *Evolutionary Biology and Human Social Organization* (Duxbury Press, Scituate, Mass., in press).

109 Pierre L. van den Berghe, "Territorial Behavior in a Natural Human Group," *Social Sciences Information*, in press.

111 The notion of the partition of the primitive world is due to Edmund Leach, "The Nature of War," *Disarmament and Arms Control* 3: 165–183 (1965).

111 Durham, William H., "Resource competition and human aggression. Part I: A Review of Primitive War," *Quarterly Review of Biology* 51: 385–415 (1976).

111 The primary source of Mundurucú warfare is Robert F. Murphy, "Intergroup Hostility and Social Cohesion," *American Anthropologist* 59: 1018–1035 (1957); and *Headhunter's Heritage: Social and Economic Change among the Mundurucú Indians* (University of California Press, Berkeley, 1960).

113 The determinants of mortality and natality schedules in hunter-gatherer peoples such as the Mundurucú are not well enough known in most cases to evaluate the density-dependent processes of population control. An excellent beginning in this important form of analysis has been made by Nancy Howell in her research on the !Kung San: see "The Population of the Dobe Area !Kung," in R. B. Lee and Irven DeVore, eds., *Kalahari Hunter-Gatherers* (Harvard University Press, Cambridge, Mass., 1976), pp. 137–

PAGE

113   151. The limited archeological evidence of the relation between population density and mode of life is reviewed in a careful manner by Mark N. Cohen in *The Food Crisis in Prehistory: Overpopulation and the Origins of Agriculture* (Yale University Press, New Haven, Connecticut, 1977).

115   Napoleon A. Chagnon, *Yanomamö: The Fierce People* (Holt, Rinehart and Winston, New York, 1968); *Studying the Yanomamö* (Holt, Rinehart and Winston, New York, 1974); and "Fission in an Amazonian Tribe," *The Sciences* 16(1): 14–18 (1976).

116   Quincy Wright, *A Study of War*, p. 100.

116   Keith F. Otterbein, *The Evolution of War* (HRAF Press, New Haven, Connecticut, 1970); and "The Anthropology of War," in J. J. Honigman, ed., *Handbook of Social and Cultural Anthropology* (Rand McNally, Chicago, 1974), pp. 923–958.

117   Andrew P. Vayda, *War in Ecological Perspective* (Plenum Press, New York, 1976).

118   The traveller's account of Maori response to firearms is from Vayda, *War in Ecological Perspective*.

119   Yanomamö quoted by John E. Pfeiffer, *Horizon*, January 1977.

120   Similar prescriptions of cross-binding ties as an aid to peace-keeping have been suggested in Margaret Mead, "Alternatives to War," in Morton Fried, Marvin Harris and Robert F. Murphy, eds., *The Anthropology of Armed Conflict and Aggression* (Natural History Press, Garden City, New York, 1968), pp. 215–218; and Donald H. Horowitz, "Ethnic Identity," in Nathan Glazer and D. Patrick Moynihan, eds., *Ethnicity: Theory and Experience* (Harvard University Press, Cambridge, Mass., 1975), pp. 111–140.

Chapter 6: Sex

122   Hereditary defects in human sex determination are described in greater detail in G. E. McClearn and J. C. DeFries, *Introduction to Behavioral Genetics* (W. H. Freeman, San Francisco, 1973); and John Money and Anke A. Ehrhardt, *Man and Woman, Boy and Girl* (Johns Hopkins University Press, Baltimore, 1972).

122   The theory of the genetic basis of sex role differences has been de-

PAGE

122 veloped by many biologists and is reviewed in detail in Wilson, *Sociobiology*, and David P. Barash, *Sociobiology and Behavior* (Elsevier, New York, 1977).

125 George P. Murdock, "World Ethnographic Sample," *American Anthropologist* 59: 664–687 (1957).

126 The relationship between polygyny and hypergamy is perceptively discussed by Pierre L. van den Berghe and David P. Barash in "Inclusive Fitness and Human Family Structure," *American Anthropologist* 79(4): 809–823 (1977).

126 Maimonides, Moses, *The Guide of the Perplexed*, translated by Shlomo Pines (University of Chicago Press, Chicago, 1963).

127 The sex differences in track performance are based on the world outdoor records to 1974 as recognized by the International Amateur Athletic Federation; the 1975 American marathon rankings were published in the *Editors of Runner's World 1975 Marathon Yearbook* (World Publications, Mountain View, California, 1976).

128 With reference to the prevalence of male dominance, see Steven Goldberg, *The Inevitability of Patriarchy* (Morrow, New York, 1973); and Marvin Harris, "Why Men Dominate Women," *New York Times Magazine*, November 13, 1977, pp. 46, 115–123.

129 The studies of sex differences in the early development of behavior are reviewed in Daniel G. Freedman, *Human Infancy*; A. F. Korner, "Neonatal Startles, Smiles, Erections and Reflex Sucks as Related to State, Sex and Individuality," *Child Development* 40: 1039–1053 (1969); and Jerome Kagan, *Change and Continuity in Infancy* (Wiley, New York, 1971).

129 Patricia Draper, "Social and Economic Constraints on Child Life among the !Kung," in Richard B. Lee and Irven DeVore, eds., *Kalahari Hunter-gatherers: Studies of the !Kung San and Their Neighbors* (Harvard University Press, Cambridge, Mass., 1976), pp. 199–217. Draper's data are few but statistically significant and in my opinion sufficient for the distinctions I have stressed in the text.

130 N. G. Blurton Jones and M. J. Konner, "Sex Differences in Behaviour of London and Bushman Children," in R. P. Michael and J. H. Crook, eds., *Comparative Ecology and Behaviour of Primates* (Academic Press, London, 1973), pp. 689–750.

PAGE

130   Eleanor E. Maccoby and Carol N. Jacklin, *The Psychology of Sex Differences* (Stanford University Press, Stanford, 1974).

130   Ronald P. Rohner, *They Love Me, They Love Me Not* (HRAF Press, New Haven, Connecticut, 1975).

131   Critical reviews of genetic and hormonal masculinization are provided in W. J. Gadpaille, "Research into the Physiology of Maleness and Femaleness," *Archives of General Psychiatry* 26: 193–211 (1972); Money and Ehrhardt, *Man and Woman*; Julianne Imperato-McGinley, Ralph E. Peterson, and Teofilo Gautier, "Gender Identity and Hermaphroditism," *Science* 191: 182 (1976); and June M. Reinisch and William G. Karow, "Prenatal Exposure to Synthetic Progestins and Estrogens: Effects on Human Development," *Archives of Sexual Behavior* 6: 257–288 (1977). The Reinisch-Karow study is especially important because it demonstrates effects on the personality of girls who were exposed prenatally to progestins but were not hermaphroditic at birth and hence not treated in any special way subsequent to birth.

134   Lionel Tiger and Joseph Shepher, *Women in the Kibbutz* (Harcourt Brace Jovanovich, New York, 1975).

134   The inhibiting influence of the deep patriarchal tradition of Israel on women's liberation is ably described by Lesley Hazleton in *Israeli Women: The Reality Behind the Myths* (Simon and Schuster, New York, 1977).

134   Hans J. Morgenthau, *Scientific Man Versus Power Politics* (University of Chicago Press, Chicago, 1946). Morgenthau eloquently stated his argument that science can have little to say concerning political behavior and matters of the spirit. For reasons expressed in the present book, I am more optimistic but do not dispute the necessity of choices beyond the reach of scientific objectivity.

135   The source of the statistics on American family structure is the Population Reference Bureau, as cited in "The family in transition," *The New York Times*, November 27, 1977, p. 1.

136   Herbert G. Gutman, *The Black Family in Slavery and Freedom 1750–1925* (Pantheon Books, New York, 1976).

136   Carol B. Stack, *All Our Kin* (Harper & Row, New York, 1974).

137   Jerome Cohen and Bernice T. Eiduson, "Changing Patterns of Child Rearing in Alternative Life-Styles," in Anthony Davids, ed.,

PAGE

137 *Child Personality and Psychopathology: Current Topics*, Vol. 3 (John Wiley, New York, 1976), pp. 25–68.

137 Rose Giallombardo, *Society of Women: A Study of a Women's Prison* (John Wiley, New York, 1966).

139 The theory of cooperative hunting in male groups and its implications for modern society is fully developed by Lionel Tiger in *Men in Groups* (Random House, New York, 1969).

140 Female sex substances in monkeys and their probable absence in human beings are reported by R. P. Michael, P. W. Bonsall, and Patricia Warner, "Human Vaginal Secretions: Volatile Fatty Acid Content," *Science* 186: 1217–1219 (1974).

143 I am grateful to Dr. John E. Boswell of Yale University for information on the world distribution of the acceptance of homosexual practice.

143 The comparison of homosexuality in animals and human beings is based on Frank A. Beach, "Cross-Species Comparisons and the Human Heritage," *Archives of Sexual Behavior* 5(3): 469–485 (1976); and F. A. Beach, ed., *Human Sexuality in Four Perspectives* (Johns Hopkins University Press, Baltimore, 1976).

145 L. L. Heston and James Shields, "Homosexuality in Twins," *Archives of General Psychiatry* 18: 149–160 (1968).

146 The role of homosexuals in hunter-gatherer and advanced societies is described by James D. Weinrich in "Human reproductive strategy" (Ph.D. thesis, Harvard University, 1976); and "Non-Reproduction and Intelligence: An Apparent Fact and One Sociobiological Explanation," *Journal of Homosexuality*, in press; and R. Reiche and M. Dannecker, "Male Homosexuality in West Germany — a Sociological Investigation," *Journal of Sex Research* 13(1): 35–53 (1977).

Chapter 7: Altruism

150 James Jones, *WWII* (Ballantine Books, New York, 1976). Similar impressions based on first-hand accounts are to be found in John Keegan's *The Face of Battle* (Viking Press, New York, 1976).

150 The account of animal altruism is taken from my article, "Human decency is animal," *New York Times Magazine*, October 12, 1975,

PAGE

150   pp. 38–50 (copyright © 1975 by the New York Times Company; reprinted by permission).

154   I owe the interpretation of the poet's acquiescence in death to Lionel Trilling's *Beyond Culture: Essays on Literature and Learning* (Viking Press, New York, 1955).

154   The rules of Nibbanic Buddhism are described by Melford Spiro in *Buddhism and Society: A Great Tradition and Its Burmese Vicissitudes* (Harper & Row, New York, 1970). A few Burmese Buddhists, it may be noted, work ultimately toward nirvana as a form of extinction, but most conceive of it as a kind of permanent paradise. I owe the examples of directed altruism in the Moslem world to Walter Kaufmann, "Selective Compassion," *The New York Times*, September 22, 1977, p. 27.

156   Much of the basic theory of kin selection and the genetic evolution of altruism was developed by William D. Hamilton. Robert L. Trivers first pointed out the importance of "reciprocal altruism" in human beings, which I have called "soft-core altruism" in the present book in the belief that this metaphor is more descriptive of the genetic basis. The theory of the evolution of altruism is reviewed in Wilson, *Sociobiology*, pp. 106–129. The implications of the juxtaposition of soft-core and hard-core altruism in human behavior was discussed in my comments on Donald T. Campbell's article, "On the Conflicts between Biological and Social Evolution and between Psychology and Moral Tradition," *American Psychologist* 30: 1103–1126 (1975); these remarks were published in *American Psychologist* 31: 370–371 (1976).

156   C. Parker, "Reciprocal Altruism in *Papio anubis*," *Nature* 265: 441–443 (1977).

159   The circumstances under which deceit is considered morally acceptable have been perceptively analyzed by Sissela Bok in *Lying: Moral Choice in Public and Private Life* (Pantheon, New York, 1978).

159   Donald T. Campbell, "On the Genetics of Altruism and the Counter-Hedonic Components in Human Culture," *Journal of Social Issues* 28(3): 21–37 (1972); and "On the Conflicts."

159   Milton M. Gordon, "Toward a General Theory of Racial and Ethnic Group Relations," in Nathan Glazer and D. Patrick Moyni-

PAGE

159   han, eds., *Ethnicity: Theory and Practice* (Harvard University
      Press, Cambridge, Mass., 1975), pp. 84–110.

160   Orlando Patterson, "Context and Choice in Ethnic Allegiance: A
      Theoretical Framework and Caribbean Case Study," in Glazer
      and Moynihan, *Ethnicity*, pp. 304–349.

162   "Director's law of public income redistribution" is due to Aaron
      Director and was elaborated by George Stigler. See the recent
      discussion in James Q. Wilson, "The Riddle of the Middle Class,"
      The Public Interest 39: 125–129 (1975).

163   Bernard Berelson and Gary A. Steiner, *Human Behavior: An In-
      ventory of Scientific Findings* (Harcourt, Brace & World, New
      York, 1964); Robert A. LeVine and Donald T. Campbell, *Ethno-
      centrism* (Wiley, New York, 1972); Nathan Glazer and D. P.
      Moynihan, eds., *Ethnicity: Theory and Practice.*

164   The account of Mother Theresa's activities is based on the article
      "Saints among Us," *Time*, December 29, 1975, pp. 47–56; and
      Malcolm Muggeridge, *Something Beautiful For God* (Harper &
      Row, New York, 1971).

165   Jesus to the Apostles, Mark 16:15–16.

165   Aleksandr I. Solzhenitsyn, *The Gulag Archipelago 1918–1956*,
      Vols. 1 and 2, translated by Thomas P. Whitney (Harper & Row,
      New York, 1973).

166   Lawrence Kohlberg, "Stage and Sequence: The Cognitive De-
      velopmental Approach to Socialization," in D. A. Goslin, ed.,
      Handbook of Socialization Theory and Research (Rand-McNally
      Co., Chicago, 1969), pp. 347–380; see also John C. Gibbs,, "Kohl-
      berg's Stages of Moral Development: A Constructive Critique,"
      Harvard Educational Review 47(1): 43–61 (1977).

Chapter 8: Religion

169   Robert A. Nisbet, *The Sociology of Emile Durkheim* (Oxford
      University Press, New York, 1974).

169   Ralph S. Solecki, "Shanidar IV, a Neanderthal Flower Burial in
      Northern Iraq," *Science* 190: 880–881 (1975).

PAGE

169   Anthony F. C. Wallace, *Religion: An Anthropological View* (Random House, New York, 1966).

169   *Logotaxis*: from the Greek *logos* (word, discourse) and *taxis* (orient, place); the term taxis is used in biology to designate the oriented movement of an organism toward a particular stimulus, as in phototaxis, an orientation toward light.

170   Sales of Billy Graham's *Angels* were reported in John A. Miles, Jr., *Zygon* 12(1): 42–71 (1977).

170   See, for example, *Objections to Astrology* (Prometheus Books, Buffalo, N.Y., 1975), a statement signed by 192 "leading scientists, including 19 Nobel winners," with articles by Bart J. Bok, "A Critical Look at Astrology," pp. 21–33, and Lawrence E. Jerome, "Astrology: Magic or Science?" pp. 37–62.

171   Friedrich W. Nietzsche, *The Genealogy of Morals*, English translation by Francis Golffing (Doubleday Anchor Books, New York, 1956).

171   For an illuminating discussion of Newton's religious beliefs and their relation to his scientific research, see Gerald Holton, "Analysis and Synthesis as Methodological Themata," in *The Scientific Imagination: Case Studies* (Cambridge University Press, Cambridge, 1977).

171   Alfred N. Whitehead, *Science and the Modern World* (Cambridge University Press, Cambridge, 1926); and *Process and Reality* (Macmillan, New York, 1929). For a recent exposition of process theology by a distinguished biologist who believes in its correctness, see Charles Birch, "What Does God Do in the World?" *Union Theological Seminary Quarterly* 30(4): 76–84 (1975).

172   Accounts of the extinction of the Tasmanian aboriginals are given in Alan Moorehead, *The Fatal Impact* (Hamish Hamilton, London, 1966) and Robert Brain, *Into the Primitive Environment* (Prentice-Hall, Englewood Cliffs, New Jersey, 1972).

176   Ernest Jones is quoted in Conrad H. Waddington, *The Ethical Animal* (Atheneum, New York, 1961).

179   This account of the significance of ritual is from Wilson, *Sociobiology*, pp. 560–562.

180   Roy A. Rappaport, *Pigs for the Ancestors: Ritual in the Ecology*

PAGE

180 *of a New Guinea People* (Yale University Press, New Haven, 1968); and "The Sacred in Human Evolution," *Annual Review of Ecology and Systematics* 2: 23–44 (1971). The latter article is an especially significant contribution to the sociobiology of religion.

181 For an excellent review of the functional analysis of witchcraft, see Robert A. LeVine, *Culture, Behavior, and Personality* (Aldine, Chicago, 1973).

181 Keith Thomas, "The Relevance of Social Anthropology to the Historical Study of English Witchcraft," in Mary Tew Douglas, ed., *Witchcraft Confessions and Accusations* (Tavistock, London, 1970), pp. 47–79. See also Keith Thomas, *Religion and the Decline of Magic* (Charles Scribner's Sons, New York, 1971); and Monica Wilson, *Religion and the Transformation of Society: A Study of Social Change in Africa* (Cambridge University Press, Cambridge, 1971).

183 John E. Pfeiffer, *The Emergence of Society: A Prehistory of the Establishment* (McGraw Hill, New York, 1977).

184 Mao Tse-tung is quoted by Alain Peyrefitte in *The Chinese*.

184 Pyatakov is quoted by Robert Conquest in *The Great Terror: Stalin's Purge of the Thirties*, revised ed. (Macmillan, New York, 1973), p. 641.

185 Ernest Becker, *The Denial of Death* (Free Press, New York, 1973).

185 Peter Marin, "The New Narcissism," *Harper's* (October 1975), pp. 45–56.

187 This translation of Num. 31: 25–30 is from *The New English Bible*, 2d ed. (Oxford University Press and Cambridge University Press, New York, 1970).

188 Hans J. Mol, *Identity and the Sacred: A Sketch for a New Social-Scientific Theory of Religion* (The Free Press, New York, 1976). Mol's conclusions are all the more interesting in that they were derived without reference to sociobiology. The evolutionary stages of religious practice have been ably traced in Robert N. Bellah, *Beyond Belief: Essays on Religion in a Post-Traditional World* (Harper & Row, New York, 1970).

189 John W. M. Whiting, "Are the Hunter-Gatherers a Cultural

PAGE

189 Type?" in Lee and DeVore, *Kalahari Hunter-Gatherers*, pp. 336–339.

190 The correlation between pastoral life and belief in an active, moral God is documented by Gerhard E. and Jean Lenski in *Human Societies* (McGraw-Hill, New York, 1970).

192 My thinking on the relation between science and religion has been greatly influenced by the writings of Robert A. Nisbet, especially his review of C. D. Darlington's *The Evolution of Man and Society* in *The New York Times Book Review*, August 2, 1970, pp. 2–3, 26; Donald T. Campbell, "On the Conflicts between Biological and Social Evolution and between Psychology and Moral Tradition," *American Psychologist* 30: 1103–1126 (1975); Ralph W. Burhoe, "The Source of Civilization in the Natural Selection of Coadapted Information in Genes and Culture," *Zygon* 11(3): 263–303 (1976); John A. Miles, Jr., "Burhoe, Barbour, Mythology, and Sociobiology," *Zygon* 12(1): 42–71 (1977); and Charles Fried, "The University as a Church and Party," *Bulletin of the American Academy of Arts and Sciences* 31(3): 29–46 (1977).

Chapter 9: Hope

197 Henry Adams, *Mont-Saint-Michel and Chartres* (Houghton-Mifflin, Boston, 1936).

198 George C. Williams, *Sex and Evolution* (Princeton University Press, Princeton, N.J., 1975).

198 Most societies are against extreme cruelty in the form of genocide, torture, forced labor, and the forced separation of families, but more refined human rights in the European-American sense still find limited acceptance. See Peter L. Berger, "Are Human Rights Universal?" *Commentary* 64: 60–63 (September 1977).

199 A few scientists have begun trying to devise methods for externalizing and making objective evaluations of secondary values. See Kenneth R. Hammond and Leonard Adelman, "Science, Values, and Human Judgment," *Science* 194: 389–396 (1976); and George E. Pugh, *The Biological Origin of Human Values* (Basic Books, New York, 1977).

PAGE

200   Love song by Sappho to Anactoria ("To an army wife in Sardis"),
translation by Mary Barnard, in *Sappho: A New Translation*
(University of California Press, Berkeley and Los Angeles, 1958;
copyright © 1958 by The Regents of the University of California
and reprinted by permission of the University of California Press).

200   For details of the dream-activation hypothesis, see Robert W.
McCarley and J. Allan Hobson, "The Neurobiological Origins of
Psychoanalytic Dream Theory," *American Journal of Psychiatry*
134: 1211–1221 (1977); and J. Allan Hobson and Robert W. Mc-
Carley, "The Brain as a Dream State Generator: An Activation-
Synthesis Hypothesis of the Dream Process," *American Journal
of Psychiatry* 134: 1335–1348 (1977).

202   This translation of *Job* 38:2–3, 16–18 is from *The New English
Bible*.

202   For a recent account of the early history of life as reconstructed
by the biochemists and paleontologists, see Robert M. Schwartz
and Margaret O. Dayhoff, "Origins of Procaryotes, Eucaryotes,
Mitochondria, and Chloroplasts," *Science* 199: 395–403 (1978).

203   This statement about the equating of intellectuals with social scientists and humanists is based on the opinion survey reported by
Charles Kadushin, "Who Are the Elite Intellectuals?" *The Public Interest* 29: 109–125 (1972).

204   I have discussed the directions in which population biology and
sociobiology must proceed in order to accommodate human behavior in "Some Central Problems of Sociobiology," *Social Sciences Information* 14(6): 5–18 (1975).

209   The translation of the passage from "Prometheus Bound" is by
David Grene and is in *Aeschylus II*, edited by David Grene and
Richmond Lattimore as part of *The Complete Greek Tragedies*
(University of Chicago Press, Chicago, 1956). Reprinted by per-
mission of The University of Chicago Press.

Index

Moynihan, D. Patrick, 240, 244–245

Muggeridge, Malcolm, 164–165, 245

Muller, H. J., 229

Mundurucú, 111–115, 239

Murder, 83, 103–104

Murdock, George P., 21–22, 227, 241

Murphy, Jane, 59, 232

Murphy, Robert F., 111, 113, 239, 240

Music, 183–184

Myth, 189–190, 200–201, 206–207

Nationalism, 92, 163–164

Natural law, 141–142

Natural selection, 38

Naturalism, 2

Navaho, 49–50

Nāyar, 135–136

Neanderthal man, 87, 169

Neisser, Ulric, 235

Neurobiology, 8

New naturalism, 8, 223

Newborn infants, 48–54

Newell, Allen, 226

Newton, Isaac, 171, 246

Nichols, Robert C., 46, 230, 234

Nietzsche, Friedrich, 171, 246

Nirvana, 154, 244

Nisbet, Robert A., 4, 245, 248

Northwest Coast Indians, 180

Nozick, Robert, 5–6, 17, 226

Nuclear family, 135, 147–148

Nuclear war, 117

Nyansongan, 182

Objectification, 188

Ona Indians, 108

Oster, George F., 228, 235

Otterbein, Keith, 116–117, 240

Paiute, 109

Palestine, 155

Parker, C., 244

Parsons, P. A., 230

Patterson, Orlando, 80–81, 160–162, 236, 245

Paul VI, 141

Peterson, Ralph E., 242

Petrinovich, Lewis F., 232

Peyrefitte, Alain, 3, 223, 247

Pfeiffer, John E., 183–184, 236–237, 240, 247

Phenylketonuria (PKU), 57–58

Phillips, John L., Jr., 234

Phobias, 68

Piaget, Jean, 66–67, 234

Pilgrim's Progress, 154

Polygyny, 125–126

Premack, David, 26–27, 227

Pribram, Karl H., 233, 235

Price, Barbara, 237

Primates, 20–21

Prisons, 137

Process theology, 171–172, 246

Prometheus, 208–209

Prostitution, 126

Psammetichus, 24

Pseudospeciation, 70

Psychology, 33

Puerto Ricans, 159

Pugh, George E., 225, 248

Pyatakov, Grigori, 184–185, 247

Index